D0301370

ST. MARY'S
COLLEGE OF EDUCATION
LIBRARY
515·8
BAL
27903

MATHEMATICAL TOPICS
General Editor: C. PLUMPTON

AN INTRODUCTION TO REAL ANALYSIS

AN INTRODUCTION TO REAL ANALYSIS

BY

DEREK G. BALL

PERGAMON PRESS

Oxford · New York · Toronto · Sydney · Braunschweig

Pergamon Press Ltd., Headington Hill Hall, Oxford
Pergamon Press Inc., Maxwell House, Fairview Park, Elmsford, New York 10523
Pergamon of Canada Ltd., 207 Queen's Quay West, Toronto 1
Pergamon Press (Aust.) Pty. Ltd., 19a Boundary Street, Rushcutters Bay, N.S.W. 2011, Australia
Vieweg & Sohn GmbH, Burgplatz 1, Braunschweig

First edition 1973
Library of Congress Catalog Card No. 72-84200

Printed in Hungary

08 016937 6 (flexicover)
08 016936 8 (hard cover)

CONTENTS

PREFACE

THIS book has been written for the use of students in colleges of education and others with similar needs. It aims to present the concepts of real analysis and to indicate the problems which necessitate the introduction of these concepts. The introductory chapter is particularly important in this respect.

A certain amount of the terminology and notation of modern algebra is used in this book, and this is explained in Chapter 1. In Chapter 2 the real numbers are built up from the natural numbers. The definitions and proofs required to give a complete account in this respect are sometimes difficult and often tedious. Consequently the only property dealt with at length is the one concerning least upper bounds. The proofs of other properties are either left as exercises or dealt with in the appendix to Chapter 2.

Throughout the book the stress has been upon concepts and interesting results rather than upon techniques, although the latter have, of course, been discussed where relevant.

The exercises in the main body of the chapters are intended to facilitate understanding of the subject-matter. Those at the end of chapters are mainly from examination papers.

I am grateful to Dr. Plumpton, for asking me to write this book for Pergamon Press and to the University of London for permission to use questions from their examination papers.

D. G. BALL

The symbol □ is used throughout the book to indicate that a proof has been concluded. The following abbreviations are used to denote that a question is taken from a London university examination paper:

T.C. Certificate in Education Examination.

B.Ed. B.Ed. Examination.

B.Sc. B.Sc. General Examination (either Part I or Part II).

INTRODUCTION

THE PURPOSE OF REAL ANALYSIS

Counting and measuring: the rationals

Counting and measuring are two fundamental activities generally associated with mathematics, and mathematics is expected to provide a system in which these activities are possible. The purpose of this introduction is to put before the reader some of the problems which are met in the use of numbers for measuring—problems which provide motivation for the creation of real analysis.

In counting we make use of the natural numbers 1, 2, 3, 4, . . ., and, to some extent, zero. These numbers, however, do not suffice for measuring. For this we frequently need to subdivide our basic unit. If we subdivide it into q parts and take p of them, we write the result p/q. Numbers like this are called fractions.

In many applications of number it is important to take account of direction as well as of magnitude. So the need for negative numbers, integers, and fractions, arises. These negative numbers, together with zero and the positive integers and fractions, give us a set of numbers called the rationals. Since by the use of rationals any given unit may be divided into as many parts as we please, the rationals are clearly sufficient for the practical purpose of expressing the result of any measurement, however accurate that measurement may be.

It is also convenient to combine rationals together in various ways to give expression to the measures of related quantities; thus we add, subtract, multiply, and divide rationals (with the proviso that we do not divide by zero). All this will be familiar to the reader.

Functions

It is often useful to relate one of the quantities being measured to another. For example, we may relate the distance through which a stone has fallen to the time which has elapsed since the start of the fall. When we relate two variable measures we sometimes find that there is a simple mathematical law connecting them. Alternatively, the law may be more complex or the relationship may be apparently lawless. We may describe relationships between variable measures mathematically by use of the language of relations and functions.

The development of the set of rationals from the set of natural numbers, the rules governing their combination, the laws satisfied by these combinations (the associative law, the commutative law, and so on), and the definitions and logical properties of relations and functions all belong to the concern of algebra. What they all have in common is that their definitions and descriptions are finite. However, as soon as we consider a theory of numbers adequate for providing a description of measure in many very familiar situations, algebra is insufficient and we are led to use infinite processes, as we shall now show.

$\sqrt{2}$

Suppose we draw an isosceles right-angled triangle whose equal sides are each 1 cm in length. We may measure the length of the hypotenuse (to some particular degree of accuracy) and express the measure as a rational, perhaps as 1.4 cm.

However, we may also calculate the length of the hypotenuse. When we do so we find, of course, that the length is $\sqrt{2}$ cm, where $\sqrt{2}$ is a positive number which, when multiplied by itself, gives 2. Now in section 2.4 we shall show that there is *no* rational number whose square is equal to 2.

This is very inconvenient. We may readily find numbers which are equal to $\sqrt{4}$ or $\sqrt{9}$ or $\sqrt{49}$, for example, and it would be a great advantage to be able to give a meaning to $\sqrt{2}$, $\sqrt{3}$, or $\sqrt{n}$, where n is any natural number. This we cannot do, at least while we are relying on rational numbers. We may find rationals whose square is nearly 2, we may find rationals whose square is as near 2 as we please, but this is not the same thing. It may be sufficient for some practical purposes, but it makes any manipulation of square roots a chancy business. It looks as though our system of rational numbers needs further extension.

Decimals

One way of attempting to evaluate $\sqrt{2}$ is by the use of decimals. What are decimals? By making use of our place-value notation, and by writing figures to the right of the units figure, we are able to give expressions for *certain* rationals. Thus $\frac{1}{2}$ may be written as 0.5, 4/25 as 0.16, and so on. But when we try to represent $\frac{1}{3}$ in this notation we find that it cannot be done. The usual division algorithm gives us 0.3333333..., but the process never ends. We may, and do, say that $1/3 = 0.33333\ldots$, and sometimes write this $0.\dot{3}$, but what exactly do we mean by this? And if we have another such expression, say $4/11 = 0.3636363636\ldots$, is it legitimate to add these expressions and obtain $23/33 = 0.69696969\ldots$? Or to multiply them? What now of $\sqrt{2}$? We find that $1.4^2 < \sqrt{2} < 1.5^2$, $1.41^2 < 2 < 1.42^2$, and so on, so that in some sense $\sqrt{2} = 1.4142\ldots$. Here, apparently, we get no repetition in the pattern of digits. So the meaning of the sequence of dots is far from clear. If we are to use decimals to express rationals like $\frac{1}{3}$ and objects like $\sqrt{2}$, we are faced with the need for using infinite sequences of digits, and what we are doing needs adequate explanation.

Area and volume

The need to consider infinite sequences arises also in quite a different context. In order to measure the area of plane sets our first task is to choose an appropriate unit. Since area is a measure of the amount of surface covered, a suitable unit for measuring area will be one which when reproduced will cover the whole plane without leaving gaps (tessellate the plane). This criterion gives us several possible units, e.g. triangles, quadrilaterals, and regular hexagons, but the one universally chosen, because it is in so many ways the most convenient, is the square. Once we have picked upon a particular square as a unit, we may conveniently give a measure to the area of rectangles (when we have surmounted one difficulty mentioned in section 8.1) by covering the rectangles with unit squares in the obvious way and then counting the squares and parts of squares used. In this way we see, for example, that the area of the rectangle in Fig. I.1 is $8\frac{1}{6}$ units.

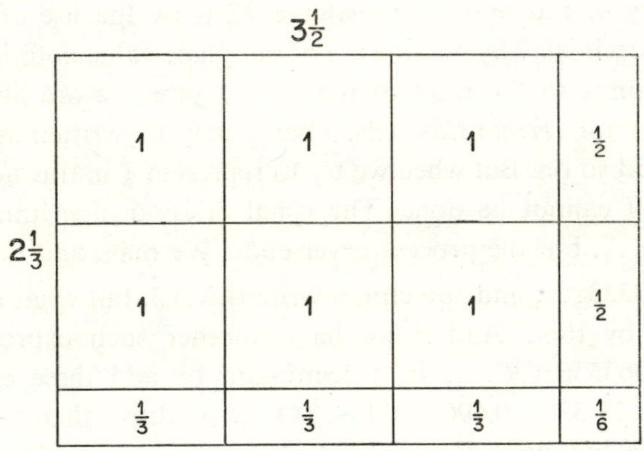

$3\frac{1}{2}$

$2\frac{1}{3}$

1	1	1	$\frac{1}{2}$
1	1	1	$\frac{1}{2}$
$\frac{1}{3}$	$\frac{1}{3}$	$\frac{1}{3}$	$\frac{1}{6}$

FIG. I.1.

By suitably rearranging a parallelogram as a rectangle we may find the area of a parallelogram, and hence of a triangle, and hence of a polygon. But what of a shape whose boundary is a curve instead of a set of straight-line segments? Suppose we wish to obtain a measure of the area of the shape shown in Fig. I.2. We may cover the shape as far as possible with unit squares, but what of the pieces round the edge? As their boundaries are still (partially) curved they are not recognizable fractions of squares. So we may ask firstly is there a number which meaningfully measures the area of the shape, and, secondly, how do we find this number for a given shape? Repeatedly subdividing our unit will give us a way of approximating to the area.

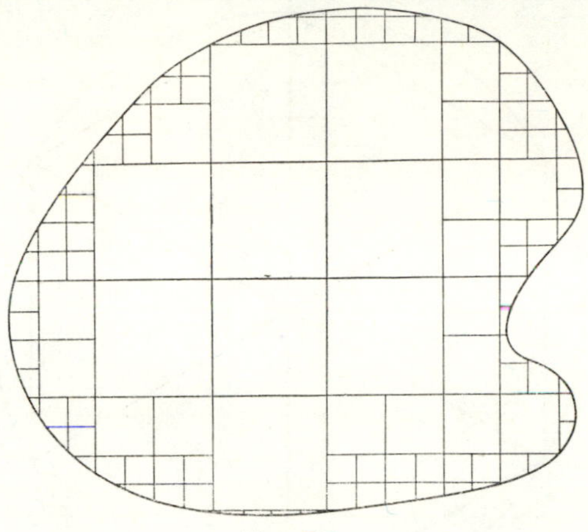

FIG. I.2.

By means of this repeated subdivision we are effectively inscribing in the shape a sequence of rectilinear polygons, each of which fills the shape more completely than its predecessor. As the process of approximation will never end, we are left with an infinite sequence of areas which we hope approaches nearer and nearer to some number which may be called the true area of the shape.

π

When we measure quantities connected with a circle, we have reason to use the ratio of its circumference to its diameter, which (since all circles are similar) is a constant π. The number π may be obtained approximately by drawing a circle and measuring its circumference and diameter. However, it is desirable to be able to calculate π. We may approximate to π by inscribing in a circle regular polygons with an increasing number of sides and calculating the perimeters of the polygons. (By inscribing a hexagon in a circle as in Fig. I.3 we see immediately that $\pi > 3$). This is a laborious process. Alternatively, we may calculate π by the use of some infinite series. For example, we show that π is given by Euler's equation

$$\pi/4 = 1-1/3+1/5-1/7+1/9-1/11+\ldots .$$

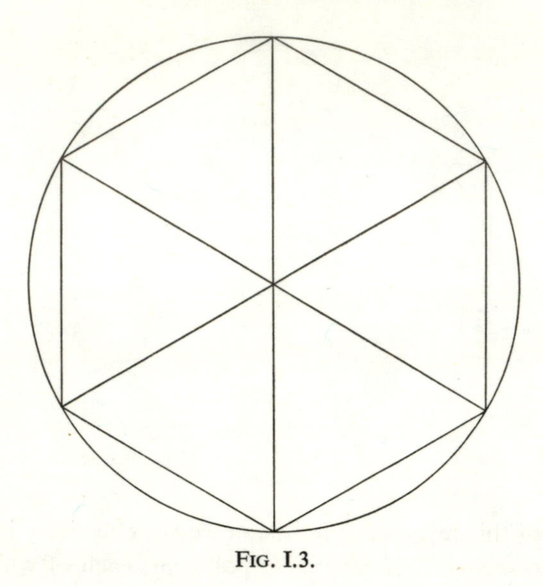

FIG. I.3.

Here we have the sum of an infinite number of terms, an infinite series. Just how do we sum an infinite number of terms; and why is π equal to the sum of this particular series?

Trigonometric functions

In the calculation of lengths and angles we often have occasion to use the trigonometric functions sine, cosine, and tangent. These functions are customarily defined in terms of the ratios of lengths of sides of right-angled triangles (or something equivalent). Thus we may find the value of the sine of 42° by drawing a right-angled triangle containing an angle of 42° and measuring two of its sides; but the accuracy of this process would not be great, so again we prefer to calculate. We may use infinite series for the evaluation of trigonometric functions. We have, in fact,

$$\sin x = x - x^3/3! + x^5/5! - x^7/7! + x^9/9! - \dots$$

provided that the angle x is measured in radians. This series may be used to evaluate $\sin x$ as accurately as we please. But, again, we are faced with the need to discuss the meaning of the sum of an infinite number of terms.

Logarithms

Logarithms give a familiar enough process for speeding approximate multiplication. We have all used the fact that $\log_{10} 2 = 0.3010\ldots$ at some time. But, again, it may be readily shown that there is no rational number x with $10^x = 2$, so that if $\log_{10} 2$ is to have meaning as a number we must again consider enlarging our concept of what is meant by a number.

In order to calculate the values of logarithms we once more make use of infinite series.

Solutions of equations: continuity

Suppose we wish to know how many real roots there are for the equation $x^2 = \cos x$, and also their approximate size. A familiar method is to draw the graphs of $y = x^2$ and of $y = \cos x$, as in Fig. I.4, and to look at their points of intersection. We deduce that in this case

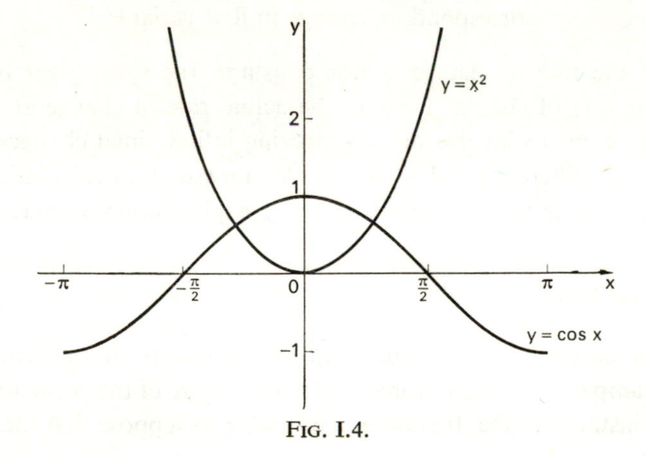

FIG. I.4.

there are exactly two real roots $\pm\alpha$, say. How do we arrive at this conclusion? We assume that $y = \cos x$ and $y = x^2$ are continuous curves (that there are no breaks in them), so that they have to cross at some point between zero and $\pi/2$ in order that $y = \cos x$ should be above $y = x^2$ at zero and below it at $\pi/2$. The point at which they cross gives us some definite number α with $\cos \alpha = \alpha^2$.

Exactly what do we mean by continuity, and how do we know when a particular curve is continuous? And does *every* point on the x-axis correspond to some number?

Rates of change

We mentioned earlier that if we have two variable quantities their measures are often related to one another and the language of functions used to describe this relationship. When we have such a situation it is often important to know that rate at which one variable is changing with respect to the other. For example, if we relate the distance travelled by a moving body to the time which has elapsed, the rate at which the distance changes with respect to the time is a measure of the velocity of the object. When the rate of change is constant it may readily be measured by the ratio

$$\frac{\text{change in second variable}}{\text{corresponding change in first variable}}.$$

But if the rate of change is not constant, the ratio gives only an average rate of change. Finding the actual rate of change at a given instant seems to involve our considering infinitesimal changes in the variables. Differential calculus provides a method for calculating rates of change at an instant. Again, however, explanation is required.

Growth functions

Suppose we have a population (of people, insects, or uranium atoms, for example) and wish to forecast the future size of the population at a given instant of time. It is often reasonable to suppose that the factors causing growth or decay affect a certain percentage of the population. The appropriate mathematical model is therefore a function of time whose rate of change is proportional to its size at any instant; exp x gives us such a function, which is conveniently defined by

$$\exp x = 1+x+x^2/2!+x^3/3!+x^4/4!+\ldots.$$

Here we have another infinite series. Furthermore, if we are to develop the properties of the exponential function from this definition we have to be able to operate confidently with infinite series.

Real analysis

All the problems mentioned above are raised because they speak of the need to introduce infinite processes into mathematics. They point to the need for a more comprehensive set of numbers, an understanding of infinite series, infinite sequences, continuity, and so on. When infinite processes were first explored, all sorts of techniques were developed in order to furnish answers to the questions raised above and many others; but the concepts that underlay the techniques and the validity or otherwise of the formal processes were not usually investigated. This is no doubt as it should be; mathematics forged ahead in the development of new processes to meet the requirements of the time.

However, in the last century mathematicians began taking a closer look at the concepts which lay behind the use of infinite processes and examining the validity of the techniques. They discarded many of the metaphysical explanations of their predecessors and substituted precise descriptions of the processes used. This examination of the concepts and investigation into the validity of the techniques of infinite processes are the concern of analysis. In this book we shall consider only the analysis of real (as opposed to complex) numbers.

BALL: An Introduction to Real Analysis

ERRATA

p. 30	lines 20 and 22	for	$b = 0$	read	$b \neq 0$
p. 47	line 22	for	$\lvert a_n \rvert b_n$	read	$\lvert a_n \rvert \lvert b_n \rvert$
p. 48	line 19	for	$\lim a_n \geqslant \lim b_n$, or $l \geqslant m$.	read	$\lim a_n \leqslant \lim b_n$, or $l \leqslant m$.
p. 56	line 14	for	$0 < x_n^2 < \sqrt{(2/n)}$.	read	$0 < x_n < \sqrt{(2/n)}$.
p. 77	line 18	for	$\lim T_n = \lim S_n$.	read	$\lim T_n = c \lim S_n$.
p. 127	line 25	for	$j(x) = f(x) + g(x)$	read	$j(x) = f(x) - g(x)$
p. 154	line 7	for	$G(x) = F(x)G(x)$	read	$G(x) = f(x)g(x)$
p. 175	lines 11 and 17	for	$h \to 0$	read	$x \to a$
p. 184	line 3	for	$f^{(n)}(a)$	read	$f^{(n)}(0)$
p. 210	line 7	for	$x \to u$	read	$x \to v$
p. 240	line 12	for	$(x_r - x_{r-1})m$	read	$(x_r - x_{r-1})M_r$
p. 243	line 9	for	$\underline{\int_a^b} f\,dx = \underline{\int_b^c} f\,dx$	read	$\underline{\int_a^b} f\,dx + \underline{\int_b^c} f\,dx$
p. 276	line 2	for	$x \to \infty$	read	$X \to \infty$
	line 8	for	$x \to -\infty$	read	$X \to -\infty$

CHAPTER 1

SETS, RELATIONS, AND FUNCTIONS

THIS first chapter is not about analysis. Its purpose is to acquaint the reader with some of the concepts and terminology of algebra with which he may be unfamiliar and which we shall use subsequently. It is particularly important that the reader should gain from this chapter a clear idea of what is meant by a function.

1.1. Sets

A *set* is the word we shall use for a collection of objects. The objects may be of any nature whatever, but they must be distinguishable one from another, and the set must be clearly defined so that we know exactly which objects belong and which do not. Sets may be finite or infinite, descriptions which are probably meaningful to the reader but which we shall define formally later in this chapter.

The objects in the set are usually called *elements* and are described as members of the set. If we denote a set by a capital letter, say A, and the elements by small letters, say $x, y, \ldots$, then we write $x \in A$ to denote that x is a member of the set A. If x is not a member of A we write $x \notin A$.

Usually a set will be put together according to some rule. We may, for example, consider the set V whose elements are the vowels in the English alphabet. It is customary to write such a set $V = \{a, e, i, o, u\}$, the set being described by listing its elements inside curly brackets. Note that by listing the elements in this way we make the definition of the set precise and in particular exclude y, which may or may not have been included under the original definition of V. We write $e \in V$ or $o \in V$ but $p \notin V, j \notin V, y \notin V$.

Or we may consider the set A whose elements are all the positive whole numbers less than 10. We write this set

$$A = \{1, 2, 3, 4, 5, 6, 7, 8, 9\} \text{ or, alternatively,}$$
$$A = \{x : x \text{ is a whole number, } 1 \leqslant x \leqslant 9\}.$$

Here the first x signifies a typical member of the set, and what follows after the colon describes the conditions imposed upon that member. If the set is large or infinite the first mode of description is inconvenient or impossible and we use the second; for example

$$B = \{x : x \text{ is an even integer, } 7 \leqslant x \leqslant 1\,000\,000\}$$

or

$$C = \{x : x \text{ is an odd integer, } x \geqslant 10\}.$$

If we have two sets A and B such that all the members of B are also members of A, we say that A contains B or that B is a *subset* of A, and write $B \subset A$. If, in addition, B is not equal to A then we say that B is a *proper subset* of A.

The set containing no members at all is called the *empty set* and is written $\{ \}$ or ϕ. We may therefore write $\phi \subset A$ for any set A.

The universal set

When we have several sets under consideration it usually happens that the elements in the various sets are all of the same kind, that is they are all drawn from some larger set, called the *universal set*. Thus the *universal set* is defined as the set which contains all the sets under consideration at a given time.

For our purposes the universal set will almost always be a set of geometric points (e.g. the set of all points in the plane) or a set of numbers (e.g. the set of integers) or a set whose members are themselves sets (e.g. the set of all subsets of $\{1, 2, 3\}$, i.e. the set

$$\{\{1, 2, 3\}, \{1, 2\}, \{1, 3\}, \{2, 3\}, \{1\}, \{2\}, \{3\}, \phi\}\}.$$

The universal set may be finite or, more usually, infinite. It is often denoted by the symbol $\mathscr{E}$.

Intersections and unions

If we have two sets A and B, both subsets of some universal set, it may happen that some elements belong both to A and B. We call the set consisting of these elements the *intersection* of A and B, and write it $A \cap B$. Thus

$$A \cap B = \{x : x \in A \quad \text{and} \quad x \in B\}.$$

If we have a finite or infinite collection of sets $D_1, D_2, D_3, \ldots,$ we may define their intersection to be the set consisting only of those elements which belong to all of them. We write

$$\bigcap_i D_i = \{x : x \in D_i \text{ for all } i\}.$$

For example, if $A = \{4, 5, 6\}$, $B = \{1, 3, 5, 7\}$, $C = \{2, 4, 6, 8\}$, then $A \cap B = \{5\}$, $A \cap C = \{4, 6\}$, $B \cap C = \phi$.

If $D_i = \{x : x < 1/i\}$ for all positive integers i, then

$$\bigcap_i D_i = \{x : x \leqslant 0\}.$$

The set consisting of all the elements which belong either to A or to B is called the *union* of A and B and is written $A \cup B$. Thus

$$A \cup B = \{x : x \in A \text{ or } x \in B \text{ or both}\}.$$

The union of a collection of sets D_i is defined as the set consisting of those elements which belong to at least one of them. We write

$$\bigcup_i D_i = \{x : x \in D_i \quad \text{for at least one value of } i\}.$$

For example, if A, B, C, and D_i are all defined as before, then

$$A \cup B = \{1, 3, 4, 5, 6, 7\}, \quad A \cup C = \{4, 5, 6, 2, 8\},$$
$$B \cup C = \{1, 2, 3, 4, 5, 6, 7, 8\}, \quad \text{and}$$
$$\bigcup_i D_i = \{x : x < 1\}.$$

If a family of sets D_i is such that the intersection of any *two* of them is the empty set, then the family of sets is said to be *disjoint*. For example, the family defined by

$$D_i = \{x : i \leqslant x < i+1\} \text{ for all integers } i$$

is disjoint.

1.2. Relations and Functions

Relations

When we are considering sets it often happens that we wish to relate the members of one set to the members of another according to some rule.

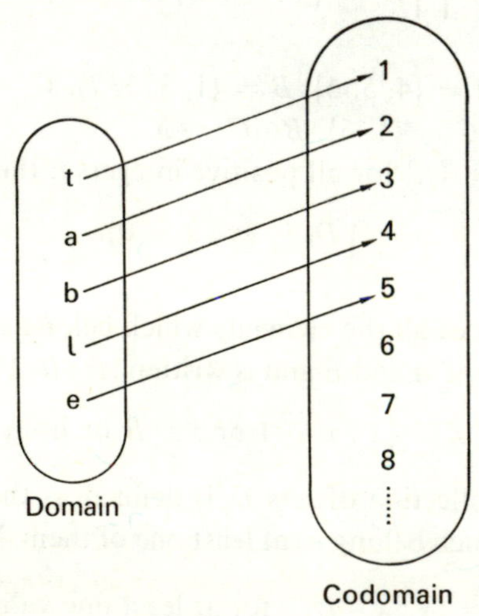

FIG. 1.1.

For example, when we count we set up such a relation. Suppose we wish to know how many letters there are in the word *table*. Then we set up a relation between the set $\{t, a, b, l, e\}$ and the set of natural numbers, as shown in Fig. 1.1.

Or again, when we consider prime factors we set up a relation between the set of natural numbers and the set of prime numbers, as shown in Fig. 1.2.

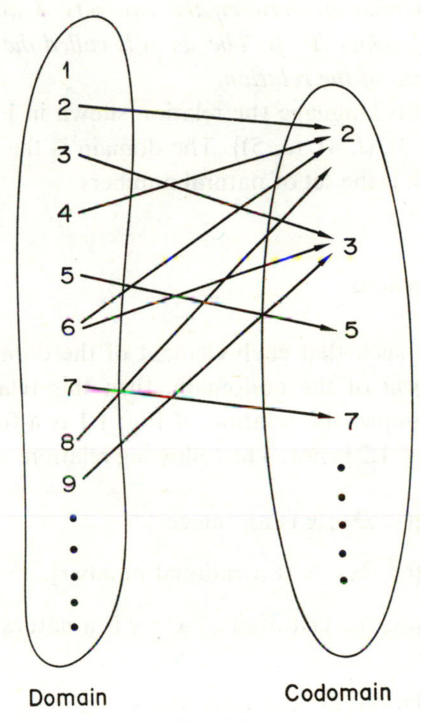

FIG. 1.2.

Each arrow in Figs. 1.1 and 1.2 is defined by two elements; an element in the first set where the arrow starts, and an element in the second where it ends. A relation may be completely defined by describing the set of arrows in these terms. Thus we make the following definitions:

DEFINITION. *The Cartesian product of two sets A and B is the set composed of all pairs of elements for which the first element of the pair is drawn from A and the second from B; it is written $A \times B$; i.e.* $A \times B = \{(a, b) : a \in A, b \in B\}$.

DEFINITION. *A* relation *between the two sets A and B is a subset of the Cartesian product $A \times B$. The set A is called the* domain *and the set B the* codomain *of the relation.*

Expressed in this language the relation shown in Fig. 1.1 is the set $\{(t, 1), (a, 2), (b, 3), (l, 4), (e, 5)\}$. The domain is the set $\{t, a, b, l, e\}$ and the codomain is the set of natural numbers.

Relations and functions

If a relation is such that each element of the domain is related to *exactly one element* of the codomain, then this relation is called a *function.* For example, the relation of Fig. 1.1 is a function, whereas the relation of Fig. 1.2 is not. The following relations are functions:

$$f = \{(x, x^2) : x \text{ is an integer}\},$$

$$g = \{(x, 3x) : x \text{ is a rational number}\},$$

$$h = \{(x, \text{the last digit of } x) : x \text{ is a natural number}\}.$$

The following relations are not functions:

$$R = \{(x, y) : x \text{ and } y \text{ are natural numbers with the same last digit}\}.$$

$$S = \{(x, y) : x \text{ and } y \text{ are integers differing by less than } 10\}.$$

It is customary to denote a relation (whether or not it is a function) by a capital letter and a function by a small letter. Also if f is a function whose domain is the set A and whose codomain is the set B, we sometimes write $f: A \to B$. If we have a function f and x is a member of the domain of f, then x is related to exactly one member of the codomain of f, denoted by $f(x)$. For example, if f, g, and h are defined as above, then $f(3) = 9$, $g(5\frac{1}{2}) = 16\frac{1}{2}$, and $h(243) = 3$.

Inverse relations and inverse functions

Suppose we have a relation R whose domain is A and whose codomain is B. Its inverse relation R^{-1} is the relation obtained simply by making B the domain, A the codomain, and reversing the directions of the arrows. Formally we have:

DEFINITION. $R^{-1} = \{(y, x) : (x, y) \in R\}$

For example, the inverse of the relation of Fig. 1.1 is shown in Fig. 1.3.

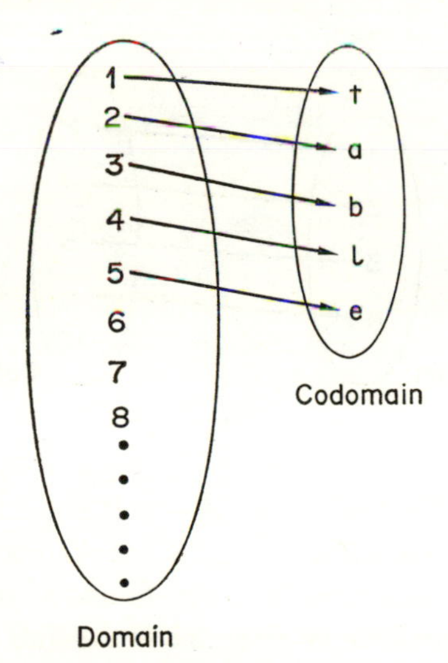

FIG. 1.3.

If the relation R is a function, then its inverse R^{-1} may or may not be. Figure 1.4 shows three functions; the inverse of (a) is a function, the inverses of (b) and (c) are not.

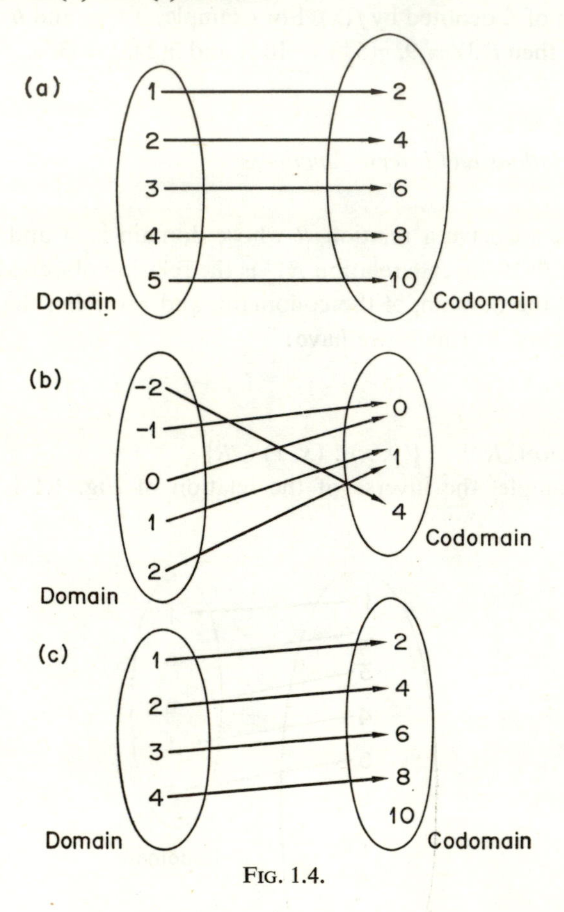

FIG. 1.4.

If the inverse of the function $f: A \to B$ is also a function, then every member of A is paired off with exactly one member of B. Under these circumstances f is described as a *one-to-one mapping*, and defines a *one-to-one correspondence* between the sets A and B. We are now in a position to give a formal definition of an infinite set.

DEFINITION. *An infinite set is a set which may be put into one-to-one correspondence with a proper subset of itself.*

For example, the function

$$f = \{(x, 2x) : x \text{ is a natural number}\}$$

defines a one-to-one correspondence between the set of natural numbers and a proper subset, the set of even numbers. Thus the set of natural numbers is infinite.

A finite set is defined to be a set which is not infinite.

Range of a function

Suppose we have a function $f: A \to B$. Since f is a function, every element of A will be related to exactly one element of B; but there may be elements of B unrelated to elements of A. Those elements of B which are related to elements of A constitute the range of f.

DEFINITION. *If we have a function $f: A \to B$, then the* range of f *is defined to be the subset of B given by $\{f(x) : x \in A\}$.*

For example, the range of the function of Fig. 1.4a is the whole of the codomain as is the range of the function of Fig. 1.4b; but the range of the function of Fig. 1.4c is the set $\{2, 4, 6, 8\}$, a proper subset of the codomain.

Composite functions

Suppose we have a function $f: A \to B$ and a function $g: B \to C$. Then we may define the composite function $h: A \to C$ by the equation

$$h(x) = g[f(x)]$$

We write $h = g \circ f$.

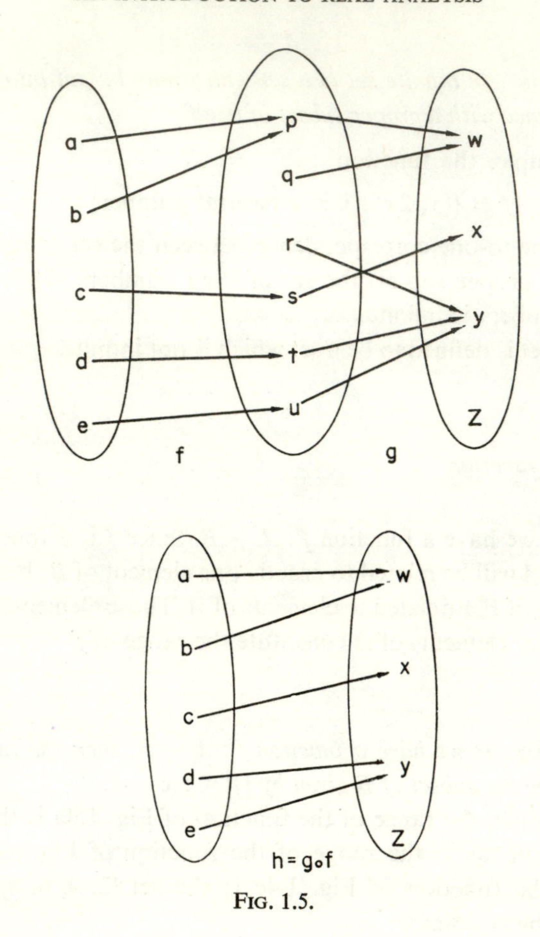

FIG. 1.5.

The relationship between f, g, and h is illustrated by Fig. 1.5. Note that $g \circ f$ means that the function f is applied first.

Equivalence relations

We now briefly define an equivalence relation and describe its properties.

DEFINITION. *If R is a relation whose domain and codomain are both the set A, then*

(i) *R is a* reflexive *relation if $(x, x) \in R$ for all $x \in A$.*

(ii) *R is a* symmetric *relation if $(y, x) \in R$ whenever $(x, y) \in R$.*

(iii) *R is a* transitive *relation if $(x, z) \in R$ whenever both (x, y) and (y, z) belong to R.*

(iv) *R is an* equivalence *relation if it is reflexive, symmetric, and transitive.*

DEFINITION. *A* partition *of a set A is a family of disjoint subsets D_i such that $\bigcup_i D_i = A$.*

DEFINITION. *If x is a member of the set A and R is an equivalence relation on A, then the equivalence class of x [which we shall write $E(x)$] is the set $\{y : (x, y) \in R\}$.*

THEOREM 1.2.1. *An equivalence relation R on the set A defines a partition of A consisting of equivalence classes.*

Proof. We have to show (i) that the equivalence classes defined by R are disjoint and (ii) that the union of all the equivalence classes is A.

(i) Suppose $E(x)$ and $E(y)$ are equivalence classes such that $E(x) \cap E(y) \neq \phi$. Let z be a member of $E(x) \cap E(y)$. Then (x, z) and (y, z) both belong to R. Therefore $(z, y) \in R$ since R is symmetric. Therefore $(x, y) \in R$ since R is transitive.

Now suppose $w \in E(y)$. Then $(y, w) \in R$. Therefore $(x, w) \in R$ since R is transitive and so $w \in E(x)$. Therefore $E(y) \subset E(x)$. So too $E(x) \subset E(y)$. Thus $E(x) = E(y)$.

(ii) Since R is reflexive $(x, x) \in R$. So $x \in E(x)$. Thus every member of A belongs to at least one equivalence class. □

Exercises for Chapter 1

1. Explain the terms *relation* and *function* and state, with reasons, which of the following are *not* functions:

(a) $\{(x, y): y = 2x+3, x \text{ and } y \text{ are integers}\}$.

(b) $\{(x, y): y^2 = 4ax, x \text{ and } y \text{ are real numbers}\}$.

(c) $\{(x, y): y = |x|, x \text{ and } y \text{ are real numbers}\}$.

(d) $\{(x, y): x \text{ is a prime factor of } y, \text{ where } x \text{ and } y \text{ are natural numbers}\}$. [T.C.]

2. Functions f and g are defined over the domain of real numbers as follows:

$$f: x \to 2x-1, \quad g: x \to x^3.$$

(a) Using the same notation express the composite functions $f \circ g$, $g \circ f$, and $f \circ f$, where $f \circ g(x) = f[g(x)]$.

(b) What is the inverse function of f and of $f \circ g$?

(c) Find any values of x which $f \circ g(x) = g \circ f(x)$. [T.C.]

3. Define an equivalence relation on a set.

State, with explanation whether, or not, each of the following relations ($\sim$) is an equivalence relation on the given set.

(a) $\sim$ means "is the brother or sister of" in the set of all human beings.

(b) $\sim$ means "has the same parents as" in the set of all human beings.

(c) $\sim$ means "is congruent to" in the set of all triangles in a plane.

(d) $\sim$ means "is less than or equal to" in the set of all real numbers

(e) $\sim$ means "has a reciprocal which, when in its lowest terms, has the same denominator as" in the set of all rational numbers. [T.C.]

CHAPTER 2

NUMBERS

NUMBERS are the basic ingredient of any theory which seeks to provide measures for mathematical situations. In this chapter we start with the natural numbers and sketch the development of the rationals from them. As indicated in the introduction the rationals are not sufficient for our purposes, so we discuss the development of a larger set, the set of real numbers. Our aim is to show the need for extension at each stage and to indicate how a logical account of the real numbers may be given. We prove the main results; however, some of the proofs of other results, which can be tedious, are omitted. The reader is referred to the appendix to this chapter if he wishes to investigate further.

The first three sections of this chapter really lie within the province of algebra. Analysis begins in section 2.4.

2.1. Natural numbers

The set of natural numbers, or counting numbers, $\{1, 2, 3, 4, 5, \ldots\}$ is the starting point for the construction of all other numbers. Natural numbers arise from the consideration of finite sets of objects. We call two finite sets similar if they are in one-to-one correspondence. We then associate a natural number with each set of similar sets. For example, the sets $\{a, e, i, o, u\}$, {cat, dog, cow, pig, horse}, $\{a, b, c, d, e\}$ are all similar, and so they are all associated with the same natural number which we call 5. We now examine some of the familiar properties of the set of natural numbers.

Order. If a set A is a subset of a set B, then the natural number associated with B is said to be greater than the natural number associated with A. Thus, given any two numbers m and n, one of the following is always true:

$$\text{(i) } m < n; \quad \text{(ii) } m = n; \quad \text{(iii) } m > n.$$

Such a set is said to be totally ordered.

The order property implies that given any finite set of natural numbers we may arrange them in order; e.g. $\{3, 7, 5, 22, 17, 4, 9, 88\}$ may be arranged in order thus:

$$3 < 4 < 5 < 7 < 9 < 17 < 22 < 88.$$

Inductive property. The natural numbers have a stronger property, however, than that of order. If we start with the number 1, which is the least natural number, we are able to write out *in order* a sequence which would ultimately include all the natural numbers. Let us clarify this statement. Obviously we cannot write down all the natural numbers since there are infinitely many of them. However, every natural number does belong to the sequence 1, 2, 3, 4, 5, . . ., in the sense that whatever number we choose we may write out the sequence until we reach that number.

This property enables us to prove theorems which are true for all natural numbers by means of the principle of mathematical induction. We assume the reader to be familiar with this method of proof.

Operations on the natural numbers. Addition of two natural numbers arises from the union of two sets of objects. It has the following important properties:

(i) The result of adding two natural numbers is another natural number. We describe this property by saying that addition is closed on the set of natural numbers.

(ii) Addition is commutative, i.e. for all m, n

$$m+n = n+m$$

(iii) Addition is associative, i.e. for all m, n, p

$$(m+n)+p = m+(n+p).$$

Multiplication of two natural numbers is defined in terms of addition. $m \cdot n$ (or simply mn) means $n+n+n+\ldots+n$, where there are m terms in the sum. Thus the product of two natural numbers is another natural number, i.e. multiplication is closed on the set of natural numbers.

3×4 means $4+4+4 = 12$ and 4×3 means $3+3+3+3 = 12$. Thus mn means something quite different from nm. However, it may be proved that the two products are always equal, i.e. that multiplication is commutative. Multiplication is also associative and distributive over addition, i.e. for all m, n, p $m\cdot(n+p) = m\cdot n+m\cdot p$.

For a set of axioms for the natural numbers, a formal definition of addition and multiplication and a sketch of how the properties mentioned here may be proved, the reader is referred to the appendix to this chapter.

Subtraction and division are defined as the inverse operations of addition and multiplication respectively. This means that if $m+n = p$ then we define $p-n$ to equal m; and if $m\cdot n = p$ we define $p\div n$ to equal m. For example, since $4+3 = 7$ we have $7-3 = 4$; and since $5\cdot7 = 35$ we have $35\div7 = 5$.

However, since there is no natural number m for which $m+8 = 2$, or for which $m\cdot4 = 17$, $2-8$ and $17\div4$ have no meaning within the set of natural numbers. Thus subtraction and division are not closed on the set of natural numbers.

It is very desirable in analysis that subtraction and division should always be possible except division by zero. In order to procure these desirable properties we augment the set of natural numbers with other numbers.

Exercises 2.1

1. Show that some of the following sets are similar, and state the natural number associated with each of them:

$\{a, b, c\}$, {Spring, Summer, Autumn, Winter}, {ace, king, queen, jack}

{Monday, Tuesday, Wednesday, Thursday, Friday, Saturday}, $\{t, u, e, s, d, a, y\}$.

2. Order the following set of numbers {3, 17, 891, 6573, 145, 61, 211}.

3. Write the following in terms of subtraction or division:

(i) $3+6 = 9$; (ii) $5 \cdot 8 = 40$; (iii) $9^2 = 81$.

4. What property of multiplication is implicit in the following method of calculating a product?

$$\begin{array}{r} 365 \\ 47 \\ \hline 2555 \\ 1460 \\ \hline 17\,155 \end{array}$$

5. Show that the set of all even numbers is inductive. Is the set of all natural numbers greater than 100 inductive? Is any subset of the natural numbers inductive?

6. Use the inductive property to show that every subset of the set of natural numbers has a least member.

2.2. Integers

So that subtraction may always be possible we add to our number set the number zero, and also a number $-m$ (which we call a negative integer) corresponding to every natural number m (which is now called a positive integer).

We may define addition and multiplication of the integers by the following rules:

(i) Addition and multiplication of positive integers is exactly as for natural numbers.

(ii) $(-m)+(-n) = -(m+n)$.

(iii) If $m > n$, $(-m)+n = n+(-m) = -(m-n)$.

(iv) If $m < n$, $(-m)+n = n+(-m) = n-m$.

(v) $m+(-m) = (-m)+m = 0$.

(vi) For any integer u, $u+0 = 0+u = u$.

(vii) $(-m)\cdot n = n\cdot(-m) = -(m\cdot n)$.

(viii) $(-m)\cdot(-n) = m\cdot n$.

(ix) For any integer u, $u\cdot 0 = 0\cdot u = 0$.

The reader will be familiar with these definitions and the fact that they give rise to a system of numbers where addition is commutative and associative, and where multiplication is commutative, associative, and distributive over addition.

As for the natural numbers, subtraction and division are defined as the inverse operations of addition and multiplication respectively. We find that, for the integers, addition, multiplication, and subtraction are always possible. Division is still more often than not impossible [e.g. we still have no meaning for $7 \div 3$ or for $(-5) \div 8$] so further extension to the number system is required.

Exercises 2.2

1. Show that the rules for addition and multiplication of integers do give in all cases an addition which is commutative and associative, and a multiplication which is commutative, associative, and distributive over addition. (Assume the corresponding properties for the natural numbers.)

2. Show that the set of integers is not inductive. Give a subset of the integers which is inductive. What condition has to be obeyed by subsets of the integers in order that they should be inductive?

2.3. Rationals

So that division (other than by zero) may always be possible we further extend the set of integers to a larger set, the set of rationals.

In order to construct the rationals we consider objects of the form p/q, where p is an integer and q is a natural number. Unfortunately we do not want all objects of this form to represent distinct numbers. For example, we require that 3/2, 6/4, 9/6, 366/244 should all represent the same number, as should $(-4)/9$, $(-8)/18$, $(-36)/81$, and so on. In order to achieve our ends we use the concept of an equivalence relation. We define a relation on the set of objects p/q by saying that p/q is related to r/s if and only if $ps = qr$. It may readily be shown that this relation is an equivalence relation, and thus defines a set of equivalence classes (see Theorem 1.2.1). For example, one such class is the set $\{3/2, 6/4, 9/6, 12/8, \ldots\}$. Each equivalence class is called a rational number; we customarily represent a rational number in a standard form by writing that member of the equivalence class for which q is least. For example, the rational $\{3/2, 6/4, 9/6, \ldots\}$ is usually written 3/2.

As is well known, we define the sum of two rationals $\{p/q, r/s, \ldots\}$ and $\{t/u, v/w, \ldots\}$ to be the rational containing the object $(pu+qt)/(qu)$ and the product of these rationals to be the rational containing the object $(pt)/(qu)$. It is necessary to show that these definitions are compatible with the equivalence relation (see exercises 2.3, question 2). With these definitions we may show that addition and multiplication have the same properties as before, that subtraction is still always possible, and that division is always possible *except by zero*. (For if a is any rational then $a \cdot 0 = 0$; thus there is no a with $a \cdot 0 = b$, and $b \div 0$ can have no meaning.) Numbers which contain an object of the form $p/1$ we call rational integers, or simply integers. Thus we may identify a subset of the rationals which behaves exactly like the integers.

Order. The rationals still retain the order property possessed by the natural numbers, since we define the rational containing p/q to be greater than the rational containing t/u if $pu-tq$ is greater than zero. (If $pu-tq$ is equal to zero the rationals are, of course, equal.) Thus any finite subset of the rationals may be arranged in order; e.g. the set $\{5/4, 6/1, 7/3, 1/2, 2/7, 1/5, 8/5\}$ may be ordered thus:

$$1/5 < 2/7 < 1/2 < 5/4 < 8/5 < 7/3 < 6/1.$$

However, the rationals do not have the inductive property, i.e. we may not write out a sequence which contains all of them arranged in order of magnitude.

Decimals. The representation of rationals by decimals is of great importance, but we are unable to give a complete account of the concepts underlying such representation until Chapter 4.

The number line; dense rationals. It is instructive to picture the rationals as points on a line, called the number line (Fig. 2.1). The method of locating rationals on the number line is illustrated in the figure and

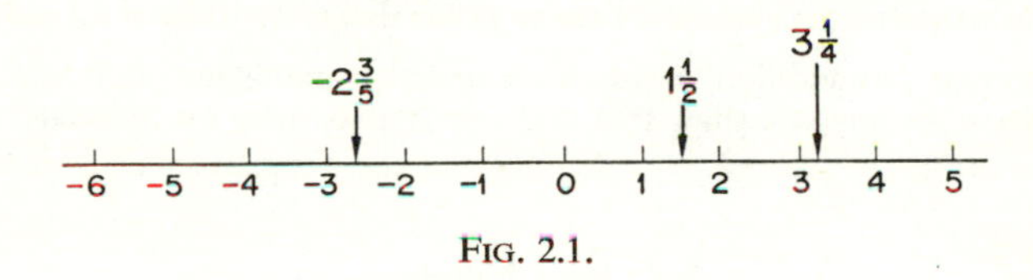

FIG. 2.1.

will seem natural to the reader. By this scheme every rational may be placed on the number line, and the rationals so placed are arranged in order of magnitude.

As is intuitively obvious from the number line, between any two rationals there is a third. For, if the rationals are p/q and r/s, then the rational $(p+r)/(q+s)$ lies between them, as the reader may show. The process may be repeated, so that between any two rationals there are infinitely many rationals. We express this fact by saying that the rationals are dense on the number line. The number line may be covered as closely as we please by rationals.

This property makes the rationals entirely adequate for expressing practical measurements. However, they are inadequate for our purposes and in need of further extension, as was indicated in the introduction.

Exercises 2.3

1. Show that the relation defined on the set of objects p/q is an equivalence relation.

2. It is necessary to show that the rule defining the sum and the product of two rationals gives the same answer no matter which object is used to represent each rational. Prove this by showing that if p/q is related to r/s, then

(i) $p/q+t/u$ is related to $r/s+t/u$, and

(ii) $p/q \cdot t/u$ is related to $r/s \cdot t/u$.

3. Show that the definition of division as the inverse operation of multiplication implies that $p/q \div r/s = (ps)/(qr)$ provided $r \neq 0$.

4. Prove that the rule defining the order property for the rationals gives the same answer no matter which object is used to represent each rational by showing that if p/q is related to r/s and $p/q < t/u$, then $r/s < t/u$.

5. Prove that the set of rationals does not have the inductive property. Is the set of positive rationals (i.e. the set of those rationals greater than zero) inductive?

6. Assuming that addition is associative and that multiplication is distributive over addition for integers, show that addition is associative for rationals.

2.4. Real Numbers

Gaps in the number line: the need to extend the rationals. In view of the fact that rationals are dense on the number line, it might seem reasonable to suppose that *all* points on the number line represent rationals. This unfortunately is not the case, as we shall now show.

Consider Fig. 2.2. A square of side 1 unit has been constructed with its diagonal along the number line. We know that the length of

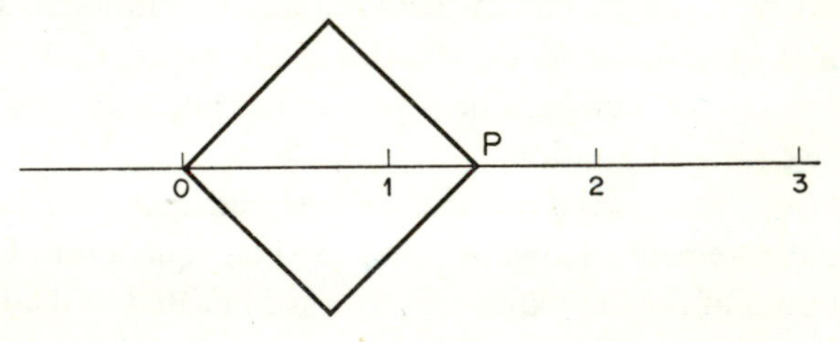

FIG. 2.2.

OP is $\sqrt{2}$ units. Thus the point P represents a rational only if $\sqrt{2}$ is rational. We shall now prove that this is not the case.

THEOREM 2.4.1. *$\sqrt{2}$ is not a rational.*

Proof. Suppose, on the contrary, that $\sqrt{2}$ is a rational and that the standard representation of this rational is p/q, so that p and q have no common factor. Then $2 = p^2/q^2$ or $2q^2 = p^2$.

Thus p^2 is even and so p is even. So let $p = 2r$, where r is a natural number. Then $2q^2 = 4r^2$, so that $q^2 = 2r^2$.

Thus q^2 is even and so q also is even. This contradicts our assumption since we have shown that p and q have the common factor 2. So $\sqrt{2}$ is irrational. □

It is very desirable that we are able to use $\sqrt{2}$ as a number. It is also desirable that every point on the number line should represent a number. We shall therefore extend the set of rationals to achieve this.

Definition of real numbers. In order to extend the rationals to overcome their deficiencies we define a larger set, the set of real numbers, effectively by associating a real number with every point of the number line. We do not use this as a definition of real numbers, however, but prefer a definition based on the consideration of sets of rational numbers.

Frequently, in real analysis, we shall use geometric intuition to suggest definitions or theorems, but we shall always define or prove without making use of the language or theorems of geometry for the following reason. As we stated in the introduction, the purpose of analysis is to clarify the concepts which underlie the techniques involved in the use of infinite processes. The axioms and definitions given by Euclid for his geometry, and normally presented in elementary geometry today, beg many of the questions which we are asking. They are particularly unsatisfactory with regard to points and lines; therefore it is probably easier to build a number system and a number theory which does not use geometrical language. Having done this it is possible to define geometry in terms of such a number theory. However, geometrical ideas often help us to have a vivid picture of what is going on. Thus we use geometry to suggest definitions and theorems, but not to define or prove.

In the next section we put into non-geometrical form the idea of defining a real number by associating it with a point of the number line.

Dedekind sections. Since we may cover the number line with rationals which are as close together as we please, between any two distinct points of the number line there is a rational. Thus any point of the number line may be distinguished from any other by knowing which rationals lie to its left and which to its right. So we define a real number by dividing the rationals into two sets L and R, L containing those rationals which lie to its left and R those rationals which lie to its right. In the definition that follows the reader should be able to see that all conditions demanded are based on geometric intuition.

DEFINITION. *A Dedekind section of the rationals is a pair of sets (L, R) whose members are all rationals and which satisfy the following conditions:*

(i) *Neither L nor R is empty.*

(ii) *Every rational belongs either to L or to R.*

(iii) *No rational belongs both to L and to R.*

(iv) *If a belongs to L and b belongs to R then $a < b$.*

(v) *L does not have a greatest member.*

Condition (v) corresponds to the geometric condition that if the point we are trying to characterize is itself a rational point, then that rational should be placed in the right-hand set. (We could, of course, have chosen to put such a rational in the left-hand set; we must, however, have a definite rule to tell us where to put it.)

DEFINITION. *A* real number *is a Dedekind section of the rationals; i.e. the set of real numbers is the set of all Dedekind sections.*

To define a particular real number we must state a rule which enables us to construct the sets L and R. For example, we may consider the particular section (L, R) defined by

$$L = \{a : a < 5/3\}, \quad \{R = a : a \geqslant 5/3\}.$$

The reader may verify that the sets L and R satisfy the condition (i)–(v) above. This section represents the real rational number 5/3. Or again we may define the section (L, R) by

$$L = \{a : a \leqslant 0, \quad \text{or} \quad a > 0 \quad \text{and} \quad a^2 < 2\}$$
$$R = \{a : a > 0 \quad \text{and} \quad a^2 > 2\}.$$

This section represents the real number $\sqrt{2}$. In this case R has no least member, so that the point which we are defining on the number line does not represent a rational.

DEFINITION. *A* real rational number *is a Dedekind section of the rationals in which R has a least member.*

Thus the rationals appear as a subset of the set of real numbers; in other words the real numbers are an extension of the rationals.

We have achieved our object of defining the real numbers without the use of geometrical language. (It may be noted in passing that we have used instead infinite sets L and R. Now there are certain logical difficulties of a sophisticated nature in the use of infinite sets, some of which may still have to be resolved. However, we cannot consider such complications here.) We now discuss the properties possessed by the set of real numbers.

Properties of the real numbers. We may define an ordering of the real numbers in the obvious way. Geometrically one real number will be greater than another if it lies to the right of the other on the number line. In terms of our definition of real numbers, we make the following definition.

DEFINITION. *Given two real numbers $x = (L_1, R_1)$ and $y = (L_2, R_2)$ we say that y is greater than x if L_1 is a proper subset of L_2. (This implies that R_2 is a proper subset of R_1.)*

If we have two real numbers (L_1, R_1) and (L_2, R_2) it may be readily shown by use of the definition of a Dedekind section that $L_1 \subset L_2$ or $L_1 = L_2$ or $L_1 \supset L_2$. Thus $x < y$ or $x = y$ or $x > y$.

If we are to use the real numbers as a generalization of the rationals, we shall require a meaning for the addition and multiplication of real numbers. Although the ideas behind the definitions are obvious, the details of relevant definitions and proofs are tedious,so we shall omit them here. The reader is referred to the appendix to this chapter however, if he wishes to repair this omission. For the moment we shall assume that addition and multiplication are defined and satisfy the usual laws and that subtraction and division (other than by zero) are the inverse operations of addition and multiplication. We shall also assume that addition and multiplication for real rationals precisely mirror the corresponding operations for rationals.

We may define the real number $\sqrt{n}$ as the Dedekind section (L, R), where

$$L = \{a : a \leqslant 0, \quad \text{or} \quad a > 0 \quad \text{and} \quad a^2 < n\},$$
$$R = \{a : a > 0 \quad \text{and} \quad a^2 \geqslant n\}.$$

It remains to be proved that $\sqrt{n}$ thus defined as the desired property $\sqrt{n} \cdot \sqrt{n} = n$. For the proof of this the reader is again referred to the appendix.

In a similar manner we may also define the real number $\sqrt[p]{n}$.

The existence of a least upper bound completeness. The set of real numbers has one very important property which is not possessed by the set of rationals. It is this property which is crucial in the use of real numbers in analysis. To describe this property we require the concept of a bounded set of numbers.

Suppose we have a non-empty set of real numbers. The set may be either finite or infinite. We may illustrate the set geometrically by marking the numbers it contains on the number line. Our set will be called bounded above if all the points lie to the left of some fixed point on the line, and it will be called bounded below if all the points lie to the right of some fixed point on the number line. Any point of the number line lying to the right of all the points in our set will be called an upper bound for the set. Similarly, any point of the number line lying to the left of all the points of our set will be called a lower bound for the set. So we make the following definitions.

DEFINITION. *The real number M is an upper bound for the set A if $M \geqslant x$ for all $x \in A$.*

DEFINITION. *The real number m is a lower bound for the set A if $m \leqslant x$ for all $x \in A$.*

DEFINITION. *The set A is said to be bounded above if it has an upper bound.*

DEFINITION. *The set A is said to be bounded below if it has a lower bound.*

DEFINITION. *The set A is said to be bounded if it is bounded above and bounded below.*

If a set is bounded then there are numbers m and M such that $m \leqslant x \leqslant M$ for all x in the set, so that all the members of the set lie on some finite line segment of the number line.

We observe that a *finite* set of real numbers is always bounded and, in addition, always has a greatest member and a least member. On the other hand, an *infinite* set of real numbers may or may not be bounded. For example, the set $\{2, 4, 6, 8, 10, \ldots\}$ is not bounded, whereas the set $\{x : 0 < x < 1\}$ is bounded. Even when an infinite set of real numbers is bounded it does not necessarily have a greatest or a least member. The set $\{x : 0 < x < 1\}$ has no greatest or least member.

Suppose now that we have a set of real numbers which we shall call A, which is bounded above. As we have remarked, A does not necessarily have a greatest member. The set A has an upper bound, say M. Now it is clear that any set which has an upper bound has more than one, in fact infinitely many. For if M is an upper bound then so is any number greater than M. There may also be upper bounds less than M. Let us consider the set of all upper bounds of the set A, and let us call this set B. As we have already remarked, B is an infinite set. Evidently B is not bounded above. It is, however, bounded below; in fact any member of A is a lower bound for B. Since B is infinite we may not immediately conclude that B has a least member. In fact B *does* always have a least member, i.e. the set A has a *least* upper bound. This we now show.

THEOREM 2.4.2. *If A is a non-empty set of real numbers which is bounded above, then from all the upper bounds we may pick a least upper bound.*

Proof. Let a typical member of the set A be $x_i = (L_i, R_i)$ so that as i varies we have all the members of A. A is bounded above, so let (L, R) be an upper bound for A.

Define L^* by $L^* = \bigcup_i L_i$ and R^* by $R^* = \bigcap_i R_i$.

We shall show that (L^*, R^*) is a Dedekind section of the rationals.

(i) Since any one of the L_i's is non-empty it is evident that L^* is non-empty.

Since (L, R) is an upper bound for A, $(L_i, R_i) \leqslant (L, R)$ for all i; so that, for all i, $R \subset R_i$. Thus $R \subset R^*$ and so R^* is non-empty.

(ii) If a rational does not belong to L^* then it does not belong to any of the L_i's. So it belongs to all of the R_i's and hence to R^*.

(iii) If a rational belongs to L^* then it belongs to at least one of the L_i's. Therefore it does *not* belong to at least one R_i. Hence it does not belong to R^*.

(iv) If a belongs to L^* and b belongs to R^*, then a belongs to at least one L_i, say to L_I. b belongs to *all* the R_i's, and so in particular it belongs to R_I. Thus $a < b$.

(v) If L^* has a greatest member then this member belongs to one of the L_i's, say to L_I. It must also be the greatest member of L_I. This contradicts the fact that (L_I, R_I) is a Dedekind section.

Thus (L^*, R^*) is a Dedekind section and represents a real number.

Since $L_i \subset L^*$ for all i, $(L_i, R_i) < (L^*, R^*)$ for all i, so that (L^*, R^*) is an upper bound for the set A.

If (L, R) is any other upper bound for A, then it is greater than all the members of A so that $L_i \subset L$ for all i.

Thus $L^* \subset L$ and so $(L^*, R^*) \leqslant (L, R)$.

Thus (L^*, R^*) is the least upper bound for the set A.

COROLLARY. *If A is a non-empty set of real numbers which is bounded below, then A has a greatest lower bound.*

Proof. Let the set A be $\{w, x, y, \ldots\}$. Consider the set $B = \{-w, -x, -y, \ldots\}$ consisting of all those real numbers which are negatives of members of A. Then B is bounded above and so has a least upper bound, say M. Then $-M$ will be the greatest lower bound for A. □

DEFINITION. *If a set of numbers is such that any of its bounded subsets has a least upper bound, the set is called* complete.

Theorem 2.4.2 proves that *the set of real numbers is complete*. The set of rational numbers, however, is *not* complete, as is demonstrated by question 7 of exercises 2.4.

Postulates for the real numbers. We may summarize the properties of the real numbers as follows:

(1) Addition and multiplication are commutative and associative, and multiplication is distributive over addition. Subtraction and division (other than by zero) are always possible. We summarize all these properties by saying that the real numbers constitute a *field*.

(2) The real numbers are *ordered*, i.e.

(i) given any two real numbers x and y exactly one of the following holds: (a) $x < y$, (b) $x = y$, (c) $x > y$.

(ii) If $x < y$ and $y < z$, then $x < z$.

(iii) If $x < y$, then $x+a < y+a$.

(iv) If $x < y$ and $a > 0$, then $ax < ay$.

(3) Every bounded subset of the real numbers has a least upper bound, i.e. the real numbers are complete.

Thus the real numbers are a complete, ordered field. In fact it may be shown that the set of real numbers is the only complete ordered field. We could, therefore, circumvent much of this chapter by defining the set of real numbers as the complete ordered field. If we did this, however, we should have to assume that such a field did exist and that it was unique.

The Archimedean property. Finally, we consider one more property of the real numbers which is intuitively obvious and which is much used in analysis.

THEOREM 2.4.3. *Given a real number $x > 0$ there is a rational a with $0 < a < x$.*

Proof. Let $x = (L, R)$. $x > 0$, so that 0 belongs to L. L has no greatest member, so there is a rational $a > 0$, belonging to L. Let

(L^*, R^*) be defined by $L^* = \{b : b < a\}$, $R^* = \{b : b \geqslant a\}$, i.e. (L^*, R^*) is the real rational a. Since a belongs to L and to R^*, $L^* \subset L$, so that $a < x$. □

THEOREM 2.4.4. (the Archimedean property). *If x and y are two real numbers with $0 < x < y$, then there exists a natural number n with $nx > y$.*

Proof. By Theorem 2.4.3 there is a rational a with $0 < a < x$. Since y is a Dedekind section of rationals there is a rational b with $b > y$. If $a = p/q$ and $b = r/s$, put $n = qr$.

Then $nx > na = (np)/q = pr \geqslant b > y$. □

Exercises 2.4.

1. Show that there is no rational whose square is (i) 3, (ii) 5.

2. If $L = \{a : a \leqslant 0$, or $a > 0$ and $a^2 < 3\}$,
and if $R = \{a : a > 0$ and $a^2 > 3\}$, show carefully that (L, R) is a Dedekind section of the rationals.

To which of L or R do the following belong?

$$-1.\ 2,\ 1\tfrac{3}{4},\ 1\tfrac{7}{10},\ 1\tfrac{29}{40},\ -1\tfrac{7}{8}.$$

3. If $L^* = \{a : a \leq 0$, or $a > 0$ and $a^2 < 5\}$, and if $R^* = \{a : a > 0$ and $a^2 > 5\}$, find a pair of rationals b, c such that $b \in L^*$, $c \in R^*$, and $c - b = 1/10$

4. If L, R, L^*, R^* are defined as in Questions 2 and 3, show carefully that $(L, R) < (L^*, R^*)$, i.e. that $\sqrt{3} < \sqrt{5}$.

5. The set A is defined by $A = \{(n-1)/n : n$ is a natural number$\}$. Show that he set A is bounded. Give three upper bounds for A. What is the least upper bound and the greatest lower bound for A?

6. Give an example of a set of real numbers which is bounded but for which neither the least upper bound nor the greatest lower bound belong to the set.

7. Show that there is no greatest rational whose square is less than 2, and no least positive rational whose square is greater than 2.

The set A consists of all rationals whose square is less than 2. Show that A is bounded but that there is no *least rational* upper bound. What is the least upper bound for A? What is the greatest lower bound for A?

8. If A is a subset of B show that the least upper bound for A is less than or equal to the least upper bound for B.

9. Why is the set of rationals not a complete ordered field? (See question 7.)

10. Why is the set of integers not a complete ordered field?

2.5. Irrationals

Those real numbers which are not rationals are called irrationals. We have already seen that $\sqrt{2}$ is irrational. So also is $\sqrt{n}$, where n is any natural number other than a perfect square; $\sqrt[3]{n}$, where n is any natural number other than a perfect cube. So also are a variety of totally different numbers like π, e, and $2^{\sqrt{2}}$, although it is sometimes very difficult to prove that a number is irrational. Purely as examples we prove two results.

(1) $\sqrt{12}$ *is irrational.* Suppose $\sqrt{12}$ is rational, so that $\sqrt{12} = p/q$, where p and q are natural numbers with no common factor. Then $p^2 = 12q^2$. Thus p^2 is divisible by 12 which is $2^2 \cdot 3$. So p is divisible by 6, and we write $p = 6r$, where r is a natural number.

Hence $36r^2 = 12q^2$ or $3r^2 = q^2$. Thus q is divisible by 3 and p and q have 3 as a common factor, which contradicts our assumption. Thus $\sqrt{12}$ is irrational.

(2) *If a and b are rationals and $b = 0$, then $a+b\sqrt{2}$ is irrational.* Suppose $a+b\sqrt{2}$ is rational. Let $a+b\sqrt{2} = c$, then $\sqrt{2} = (c-a)/b$ since $b = 0$. But if a, b, and c are rational, then so is $(c-a)/b$. Thus $\sqrt{2}$ is rational. This gives us a contradiction. Thus $a+b\sqrt{2}$ is irrational.

The representation of irrationals by decimals is an interesting and important topic; but, like the representation of rationals by decimals, it will have to wait until Chapter 4.

Exercises 2.5

1. Show that the following numbers are irrational:

(i) $\sqrt{3}$; (ii) $\sqrt[3]{2}$; (iii) $\sqrt{18}$; (iv) $\sqrt{(7/5)}$.

2. Show that $4+5\sqrt{3}$ is irrational.

3. Show that $a+b\sqrt{5}$ is irrational if a and b are rational and $b \neq 0$.

4. Show that $\log_{10} 2$ is irrational, — i.e. that there is no rational $a = p/q$ with $10^a = 2$.

5. Show that there exists a pair of positive numbers, both of which are irrational, but whose sum and product are both rational.

6. Show that $a+b\sqrt{2}+c\sqrt{3}$ is irrational if a, b, and c are rational and at least one of b and c is non-zero.

7. Prove that between any two rational numbers there is an irrational number.

8. Prove that between any two real numbers there is a rational number. (This is an extension of Theorem 2.4.3.)

2.6. Appendix

At the end of all the other chapters of this book there is a set of miscellaneous exercises. For this chapter, however, we are giving exercises which show how some of the gaps we left while we were developing the real numbers may be filled.

Peano's axioms. We may build the properties of the natural numbers from the following set of axioms, called Peano's axioms.

The set of natural numbers is a set N with the following properties:

(i) For every number $n \in N$ there is another unique number called the successor of n. [We shall write this number $s(n)$.]

(ii) There is one and only one natural number which is not the successor of any number. (We shall call this 1.)

(iii) Two different numbers do not have the same successor (i.e. if $s(m) = s(n)$, then $m = n$).

(iv) *The principle of induction.* If A is a subset of N with the two properties (a) $1 \in A$, (b) if $n \in A$ then so does $s(n)$, then $A = N$.

From these axioms we may define addition and multiplication of natural numbers and show that they are commutative and associative and that multiplication is distributive over addition. The definitions are given and the reader is left to prove the properties, although the order in which they should be proved is indicated.

DEFINITION. *We define $n+m$ by induction on m.*

Firstly we define $n+1 = s(n)$. Now, assuming that $n+m$ is defined, we define $n+s(m) = s(n+m)$.

[Note how we use axiom (iv) to establish that this definition is meaningful for all m. Firstly, $n+1$ is defined so addition is defined when $m = 1$. It is also defined for $s(m)$ whenever it is defined for m. Thus it is defined on a set A satisfying conditions (a) and (b) of (iv). So it is defined for all of N.]

DEFINITION. *We define $m \cdot n$ by induction on m.*

Firstly we define $1 \cdot n = n$. Now, assuming that $m \cdot n$ is defined, we define $s(m) \cdot n = m \cdot n + n$.

Exercises

1. Show from the definition of addition that addition is associative, i.e. that $(m+n)+p = m+(n+p)$ by induction on p. [That is, show that $(m+n)+1 = m+ +(n+1)$; and then, assuming that $(m+n)+p = m+(n+p)$, prove that $(m+n)+ +s(p) = m+\{n+s(p)\}$.]

2. Prove by induction on m that $s(n)+m = s(n+m)$ and hence prove that $n+m = m+n$.

3. Using the definition of multiplication and the result of question 2, prove that $(m+1)(p+q) = (m+1)p+(m+1)q$.

Hence prove that $m(p+q) = mp+mq$ by induction on m.

4. Show by induction that $n \cdot 1 = n$.

Hence, with the help of the result of question 3, show that $m \cdot n = n \cdot m$ by induction on m.

5. Deduce from the results of the previous two questions that

$$(p+q)m = pm+qm.$$

Hence prove by induction on p that $p \cdot (q \cdot r) = (p \cdot q) \cdot r$.

Addition and multiplication of real numbers

DEFINITION. *If* $x = (L, R)$ *and* $y = (L^*, R^*)$ *we define* $x+y = (L', R')$ *where* $L' = \{c : c = a+b,\ a \in L,\ \text{and}\ b \in L^*\}$, and R' = *the set of rationals not belonging to* L'.

Before we are able to accept this as a definition of addition of real numbers, we first have to show that (L', R') is a Dedekind section.

THEOREM. (L', R') *as defined in the definition above is a Dedekind section.*

Proof. (i) Suppose that $p \in R$ and $q \in R^*$. Then, if $p+q = r+s$, either $r \geqslant p$ or $s \geqslant q$. So we cannot find an r in L and an s in L^* with $p+q = r+s$. Therefore $p+q$ belongs to R' and so R' is non-empty. It is evident that L' is non-empty.

(ii) and (iii). That every rational belongs either to L' or to R' and that no rational belongs to both sets is immediate from the definition of R'.

(iv) Suppose that $d \in L'$ and that $e < d$. Then $d = a+b$ for some a and b, with $a \in L$ and $b \in L^*$. $e = d+e-d = (a+e-d)+b$. But $a+e-d < a$, so that $(a+e-d) \in L$. Thus $e \in L'$. Hence if $d \in L'$ and $f \in R'$ then $d < f$.

(v) Suppose L' has a greatest member c. Then $c = a+b$, where a is the greatest member of L, b is the greatest member of L^*. This contradicts the definition of L and L^*. Thus L' does not have a greatest member. So (L', R') is a Dedekind section. □

If the rationals are to be regarded as a subset of the reals, it is necessary not only that to every rational a there should be a corresponding real rational a^*, but also that the addition and multiplication of real rationals should mirror the addition and multiplication of rationals. We prove the relevant theorem for addition. The corresponding result for multiplication is left as an exercise.

THEOREM. *If* a^* *and* b^* *are real rationals corresponding to the rationals* a *and* b, *then* $a^*+b^* = (a+b)^*$; *i.e.,* a^*+b^* *is the real rational corresponding to the rational* $a+b$.

Proof. If $a^* = (L, R)$ and $b^* = (L^*, R^*)$, then the least member of R is a and the least member of R^* is b. So that, if $a^*+b^* = (L', R')$, the definition of addition gives the least member of R' as $a+b$. Thus (L', R') is the real number corresponding to the rational $a+b$. □

In view of this theorem and the corresponding theorem for multiplication, real rationals behave in precisely the same way as rationals and generally no confusion is caused if we fail to distinguish one from the other. Thus we shall write a to mean either the rational a or the real rational a^*.

DEFINITION. (the product of positive real numbers). *If x and y are real numbers with $x \geqslant 0$ and $y \geqslant 0$, we define xy thus: If $x = (L, R)$ and $y = (L^*, R^*)$, then $xy = (L', R')$, where $L' = \{c : c < 0$ or $c = ab$, $a \in L$ and $b \in L^*$, $a \geqslant 0$, and $b \geqslant 0\}$, and R' = the set of rationals not belonging to L'.*

The proof that this definition does give a Dedekind section (L', R') is left as an exercise.

DEFINITION (the real number $-x$). *If $x = (L, R)$, then we define $-x$ by $-x = (L^*, R^*)$ where $L^* = \{a : -a \in R$ and $-a$ is not the least member of $R\}$, and $R^* = \{a : -a \in L$ or $-a$ is the least member of $R\}$.*

This rather clumsy definition is necessary to ensure that L^* does not have a greatest member. The proof that the other properties of a Dedekind section are satisfied by (L^*, R^*) is left as an exercise.

DEFINITION (the product of x and y if at least one of x and y is negative). *If x is positive and y is negative then x and $-y$ are positive so that $x \cdot -y$ is defined. We define $xy = -(x \cdot -y)$. So, too, if x is negative and y is positive, we define $xy = -(-x \cdot y)$ and, if x and y are both negative, we define $xy = -x \cdot -y$.*

Exercises 2.6

1. If x, y, and z are real numbers show

(i) $x+y = y+x$; (ii) $(x+y)+z = x+(y+z)$.

2. If x, y, and z are real numbers and $x < y$, show that $x+z < y+z$.

3. By considering the case $x = \sqrt{2}$, $y = -\sqrt{2}$, explain why the following is not a satisfactory definition of the sum of two real numbers.

If $x = (L, R)$ *and* $y = (L^*, R^*)$, *then* $x+y = (L', R')$, *where*

$L' = \{c : c = a+b, a \in L$, *and* $b \in L^*\}$ *and*

$R' = \{c : c = a+b, a \in R$, *and* $b \in R^*\}$.

4. Prove that (L', R') in the definition of the product of positive real numbers is a Dedekind section.

5. Prove that (L^*, R^*) in the definition of the real number $-x$ is a Dedekind section.

6. Prove that the product of positive real numbers is associative, commutative, and distributive over addition.

7. If a^* and b^* are the real rationals corresponding to the rationals a and b and if $(a \cdot b)^*$ is the real rational corresponding to $a \cdot b$, show that $a^* \cdot b^* = (a \cdot b)^*$.,

8. Show that if c is a rational and $c < 2$, then there is a perfect square d (i.e. $d = e^2$, where e is a rational) with $c < d < 2$.
Hence show that, if $\sqrt{2}$ is defined as on p. 23, then $\sqrt{2} \cdot \sqrt{2} = 2$.

9. Explain why the following is not a satisfactory definition of the product of two real numbers:

If $x = (L, R)$ *and* $y = (L^*, R^*)$, *then* $xy = (L', R')$ *where* $L' = \{c : c = a+b,$ $a \in L$, *and* $b \in L^*\}$ *and*

R' = *the set of rationals not belonging to* L'.

CHAPTER 3

SEQUENCES

3.1. Introduction

IN THE introductory chapter we saw that some quantities may only be measured by means of a sequence of approximations. If a sequence of numbers is approximating ever more closely to some fixed real number, we call this number the limit of the sequence. In this chapter we shall examine exactly what is meant by a limit and the conditions under which sequences have limits; we shall also derive the limits of many important sequences.

A sequence is a list of real numbers arranged like the list of natural numbers. The numbers which make up the sequence may all be different, but need not be. Thus

$$1, 2, 3, 4, 5, 6, \ldots,$$
$$1, \sqrt{2}, \sqrt{3}, 2, \sqrt{5}, \sqrt{6}, \sqrt{7}, \sqrt{8}, 3, \ldots,$$
$$1, 1/2, 1/3, 1/4, 1/5, 1/6, \cdots,$$
$$1, -1, 1, -1, 1, -1, 1, \ldots, \text{ are all sequences.}$$

Since we talk about the first term, second term, third term of a sequence and so on, it is natural to make the following definition.

DEFINITION. *A* sequence *is a function whose domain is the set of natural numbers and whose range is some subset of the real numbers.*

The sequences above are defined by the functions $f(n)=n$, $f(n) = \sqrt{n}$ $f(n) = 1/n$, $f(n) = (-1)^{n+1}$ respectively. It is common practice to write the nth term of a sequence a_n instead of $f(n)$; in this notation the definition of the last sequence would read $a_n = (-1)^{n+1}$. If the nth term of a sequence is a_n we shall write the sequence as a whole $\{a_n\}$.

It is sometimes convenient to call the first term of a sequence something other than a_1. Thus we shall sometimes name our sequences by beginning with some other index number, e.g.

$$a_0, a_1, a_2, a_3, a_4, a_5, \ldots, \quad \text{or}$$
$$a_2, a_3, a_4, a_5, a_6, a_7, \ldots.$$

We note that our definition implies that all sequences are infinite in extent. This in turn means that, although sequences may not begin by following any very obvious pattern, they must eventually follow some rule or rules, or else we would be unable to define them. For example, the sequence

$$5, -7, -13, 4, -8, 16, -32, 64, \ldots,$$

begins erratically, but eventually obeys the rule

$$a_n = (-2)^{n-2}.$$

Increasing and decreasing sequences. Sequences may be classified in various ways. Here we discuss one of the most obvious and most important ways.

DEFINITION. *An* increasing sequence $\{a_n\}$ *is a sequence for which* $a_{n+1} \geqslant a_n$ *for all n.*

DEFINITION. *A* strictly increasing sequence $\{a_n\}$ *is a sequence for which* $a_{n+1} > a_n$ *for all n.*

DEFINITION. *A* decreasing sequence $\{a_n\}$ *is a sequence for which* $a_{n+1} \leqslant a_n$ *for all n.*

DEFINITION. *A* strictly decreasing sequence $\{a_n\}$ *is a sequence for which* $a_{n+1} < a_n$ *for all n.*

DEFINITION. *A* monotonic sequence *is a sequence which is either increasing or decreasing.*

For example, the following sequences are strictly increasing:

$$2, 4, 8, 16, 32, \ldots,$$

$$2, 3, 5, 7, 11, 13, 17, 19, \ldots,$$

$$4, 6, 7, 7\tfrac{1}{2}, 7\tfrac{3}{4}, \ldots,$$

$$-8, -4, -2, -1, -\tfrac{1}{2}, -\tfrac{1}{4}, \ldots;$$

the following are increasing, but not strictly:

$$1, 1, 2, 2, 4, 4, 8, 8, 16, 16, 32, 32, \ldots,$$

$$1, 2, 2, 3, 3, 3, 4, 4, 4, 4, 5, 5, 5, 5, 5, \ldots;$$

the following is neither increasing nor decreasing:

$$1, -1, 1, -1, 1, -1, 1, -1, 1, \ldots.$$

Exercises 3.1

1. Formally define the following sequences by giving the term a_n (e.g. if the sequence is 1, 4, 9, 16, ..., $a_n = n^2$):

(i) $1, 8, 27, 64, 125\ldots$

(ii) $1, 1/2, 1/4, 1/8, 1/16\ldots$

(iii) $1, -2, 3, -4, 5, -6\ldots$

(iv) $1, 1\frac{1}{2}, 1\frac{3}{4}, 1\frac{7}{8}. \ldots$

(v) $1, 0, 3, 0, 5, 0, 7, 0\ldots$

(vi) $1, -1, 0, 1, -1, 0, 1, -1, 0\ldots$

(vii) $2, 1\frac{1}{2}, 3\frac{1}{3}, 3\frac{3}{4}, 5\frac{1}{5}, 5\frac{5}{6}, 7\frac{1}{7}, 7\frac{7}{8}\ldots$

2. Say of the following sequences which are increasing, which are decreasing, and which increase or decrease strictly:

(i) $1, 8, 27, 64, 125\ldots$

(ii) $3, 7, 9, 10\frac{1}{2}, 10\frac{3}{4}\ldots$

(iii) $-100, -99\frac{1}{2}, -99\frac{1}{4}, -99\frac{1}{8}\ldots$

(iv) $100, 50, 25, 12\frac{1}{2}, 6\frac{1}{4}\ldots$

(v) $-2, 4, -6, 8, -10, 12\ldots$

(vi) $-2, -4, -6, -8, -10, -12\ldots$

(vii) $-2, -1, -4, -3, -6, -5, -8, -7\ldots$

(viii) $1, 3, 3, 5, 5, 5, 7, 7, 7, 7\ldots$

(ix) $-1, -1, -1, -1, -1, -1, -1$

(x) $5, 5, 4, 4, 3, 3, 2, 2, 1, 1, 0, 0, -1, -1\ldots.$

3. $\{a_n\}$ and $\{b_n\}$ are decreasing sequences and the sequences $\{c_n\}$ and $\{d_n\}$ are defined by $c_n = a_n + b_n$, $d_n = a_n b_n$.

(i) Is $\{c_n\}$ necessarily decreasing?

(ii) Is $\{d_n\}$ necessarily decreasing?

Justify your answers.

3.2. Limits of sequences

Sequences generally arise as sets of closer and closer approximations to some real number. The limit of a sequence is a formulation of this idea.

Consider the sequences

$$1, 4, 9, 16, 25, 36, \ldots,$$
$$2, 6, 4, 8, 10, 8, 12, \ldots .$$

The first is an increasing sequence; the second is not. However, the sequences have something in common; their terms eventually become very large. To be more precise, their terms eventually become *and stay* as large as we please. For example, if we choose the number 10 000 then eventually the terms of both sequences will become and stay greater than 10 000. The same is true whatever number we choose. We describe such a sequence by saying that it tends to infinity.

DEFINITION. *A sequence* $\{a_n\}$ *tends to infinity as n tends to infinity (written* $a \to \infty$ *as* $n \to \infty$*) if, given* any *real number K, there exists a natural number N (which generally depends upon the choice of K) such that* $a_n > K$ *for all* $n \geqslant N$.

If $a_n \to \infty$ as $n \to \infty$ we also write $\lim\limits_{n \to \infty} a_n = \infty$ or even simply $\lim a_n = \infty$. The reader should note carefully that our use of the symbol ∞ does not imply the existence of a number called infinity. The symbol is only to be used in conjunction with the idea of a limit, to have precisely the meaning described by the definition.

It is, of course, important to be able to use the definition above to show that a particular series tends to infinity. For the moment we give one easy example to illustrate this.

Consider the sequence

$$1, 4, 9, 16, 25, 36, 49, 64, \ldots.$$

Here $a_n = n^2$. If we are given $K = 1000$ we may choose $N = 32$, and then $a_n \geqslant a_{32} = 1024 > 1000$ whenever $n \geqslant 32$.

If we are given $K = 1\,000\,000$ we may choose $N = 1001$, and then $a_n \geqslant a_{1001} = 1\,002\,001$ whenever $n > 1001$.

More generally, given K, we may choose any $N > K$.

Then $a_n = n^2 \geqslant N^2 \geqslant N > K$ whenever $n > N$. Thus $a_n \to \infty$ as $n \to \infty$. Note that our eventual choice of N was (for simplicity) much larger than it need have been for most values of K (we could have chosen $N > \sqrt{K}$). This, however, does not matter. We do not necessarily have to find the smallest value of N which will do what is required, simply a value.

Consider now the sequence

$$2, -4, 8, -16, 32, -64, \ldots.$$

Here $a_n = -(-2)^n$. This sequence is not bounded above, but it does not tend to infinity, as we may readily show from the definition.

For suppose we are given $K = 100$. Since every alternate term of the sequence is negative, no matter how large we choose N we may never ensure that a_n is positive for all $n \geqslant N$, let alone greater than 100. Thus the sequence fails to satisfy the definition and does not tend to ∞.

Sequences which tend to minus infinity. We apply much the same considerations to sequences whose terms eventually become and stay very large and negative; e.g. the sequence

$$-2, -4, -6, -8, -10, -12, -14, \ldots.$$

DEFINITION. *A sequence $\{a_n\}$ tends to minus infinity as n tends to infinity (written $a_n \to -\infty$ as $n \to \infty$) if, given* any *real number K, there exists a natural number N (depending on the choice of K) such that $a_n < K$ for all $n \geqslant N$.*

If $a_n \to -\infty$ as $n \to \infty$, we also write $\lim\limits_{n\to\infty} a_n = -\infty$ *or even simply* $\lim a_n = -\infty$.

The modulus of a real number. In discussing further the concept of a limit the notion of the modulus of a real number is useful. The reader is probably familiar with this.

DEFINITION. *The modulus of the real number x, written $|x|$, is defined by*

$$|x| = x \text{ if } x \geqslant 0, \quad |x| = -x \text{ if } x < 0.$$

Thus the modulus of a real number is always greater than or equal to zero. Geometrically $|x|$ is the distance of x from zero on the number line, and $|x-y|$ is the distance between x and y on the number line. The following theorems may be easily proved by the enumeration of the different cases which arise when some or all of the numbers involved are positive or negative.

THEOREM 3.2.1. *If x and y are real numbers,* $|x+y| \leqslant |x|+|y|$.

COROLLARY. *If $x_1, x_2, x_3, \ldots, x_n$ are real numbers, then*

$$|x_1+x_2+x_3+\ldots+x_n| = |x_1|+|x_2|+|x_3|+\ldots+|x_n|.$$

THEOREM 3.2.2. *If $x_1, x_2, x_3, \ldots, x_n$ are real numbers, then*

$$|x_1x_2x_3x \ldots x_n| = |x_1|\,|x_2|\,|x_3| \ldots |x_n|.$$

THEOREM 3.2.3. *If x is a real number* and $x \neq 0$

$$\left|\frac{1}{x}\right| = \frac{1}{|x|}.$$

Convergent sequences. We now consider sequences which approximate closer and closer to some real number. Consider the sequence

$$32, 48, 56, 60, 62, 63, 63\tfrac{1}{2}, 63\tfrac{3}{4}, \ldots .$$

For this sequence $a_n = 64(1-(\frac{1}{2})^n)$. This sequence seems to be approximating ever more closely to 64. By this we mean that the terms of the sequence eventually become and stay as close to 64 as we please, or to put this another way, that the difference between the nth term and 64 is eventually as small as we please. We achieve this concept by making the following definition.

DEFINITION. *A sequence $\{a_n\}$ tends to the limit l as n tends to infinity (written $a_n \to l$ as $n \to \infty$), if, given any real number $\varepsilon > 0$, there exists a natural number N (depending on the choice of ε) such that $|a_n - l| < \varepsilon$ for all $n \geqslant N$.*

If $a_n \to l$ as $n \to \infty$, we also write $\lim_{n\to\infty} a_n = l$ *or even simply* $\lim a_n = l$.

DEFINITION. *A sequence which tends to a finite limit is said to be convergent.*

We shall later be able to prove that the sequence

$$32, 48, 56, 62, 63, 63\tfrac{1}{2}, 63\tfrac{3}{4}, \ldots,$$

satisfies the definition and tends to 64.

Our definition does *not* demand that for a sequence to tend to a limit l, each term of the sequence must be a better approximation to l

than the previous term. For example, we may show that the sequence

$$1/2, 1/2, 1/4, 1/3, 1/8, 1/4, 1/16, 1/5, 1/32, 1/6, 1/64, 1/7, \ldots,$$

tends to zero.

The sequence

$$4, 0, 6, 0, 7, 0, 7\tfrac{1}{2}, 0, 7\tfrac{3}{4}, 0, \ldots,$$

does not tend to 8, since although we can find members as close as we please to 8, we cannot find a term in the sequence such that *all* terms beyond this term are as close as we please to 8.

To summarize, a sequence may do one of four things:

(i) it may tend to infinity;

(ii) it may tend to minus infinity;

(iii) it may tend to a finite limit;

(iv) it may not have a limit at all.

DEFINITION. *A sequence which has no limit at all is said to* oscillate.

The limits of some commonly occurring sequences. Some sequences occur frequently; it will be of value both to know what their limits are and, also, to see how the definitions are employed in establishing the behaviour of a sequence.

THEOREM 3.2.4. (i) $\lim 1/n = 0$. (ii) $\lim c = c$, *where c is some real constant.* (iii) *If* $0 < |x| < 1$, $\lim x^n = 0$. (iv) *If* $x > 1$, $\lim x^n = \infty$. (v) *If p is a natural number*, $\lim n^p = \infty$. (vi) *If p is a natural number*, $\lim n^{1/p} = \infty$.

Proof. (i) By Theorem 2.4.3, given any real number $\varepsilon > 0$, there is a rational M/N satisfying $0 < M/N < \varepsilon$. Therefore for all $n \geqslant N$,

$$\left|\frac{1}{n} - 0\right| = \frac{1}{n} \leqslant \frac{1}{N} \leqslant \frac{M}{N} < \varepsilon.$$

Thus $1/n \to 0$ as $n \to \infty$.

(ii) Here every term of the sequence is equal to the number c. Given $\varepsilon > 0$, we may choose N to be any natural number we please. Then, for all $n \geqslant N$,

$$|c-c| = 0 < \varepsilon.$$

(iii) If $0 < |x| < 1$, then $|x| = 1/(1+y)$ for some $y > 0$. Therefore by the binomial theorem

$$|x|^n = \frac{1}{(1+y)^n} \leqslant \frac{1}{1+ny} < \frac{1}{ny}.$$

Given $\varepsilon > 0$, by Theorem 2.4.3 there is a rational M/N satisfying $0 < M/N < \varepsilon y$. Thus

$$0 < \frac{1}{N} \leqslant \frac{M}{N} < \varepsilon y \quad \text{and so} \quad \frac{1}{Ny} < \varepsilon.$$

So for all $n \geqslant N$,

$$|x^n - 0| = |x^n| = |x|^n < \frac{1}{ny} \leqslant \frac{1}{Ny} < \varepsilon.$$

Therefore $x^n \to 0$ as $n \to \infty$.

(iv) If $x > 1$, then $x = 1+y$ for some $y > 0$. So by the binomial theorem $x^n = (1+y)^n \geqslant 1+ny$, for all n.

Given any real number K, there is a natural number $N > K/y$. Then for all $n \geqslant N$,

$$x^n \geqslant 1+ny > Ny > K.$$

Therefore $x^n \to \infty$ as $n \to \infty$.

(v) Given any number K there is a natural number $N > K$. Since p is a natural number $n^p \geqslant n$ for all natural numbers n. Thus for all $n \geqslant N$, $n^p \geqslant n \geqslant N > K$. Thus $n^p \to \infty$ as $n \to \infty$.

(vi) Given any number K there is a natural number $N > K^p$. Thus for all $n \geqslant N$, $n^{1/p} \geqslant N^{1/p} > K$. Thus $n^{1/p} \to \infty$ as $n \to \infty$. □

Exercises 3.2

1. State the nth term and the limit of the sequence

$$3, 2\tfrac{1}{3}, 2\tfrac{1}{9}, 2\tfrac{1}{27}, 2\tfrac{1}{81} \ldots.$$

Given $\varepsilon = 1/1000$, what is the least value which N may have in order to satisfy the requirements of the definition of a limit?

2. State the limits of the following sequences Justify your answers carefully from the relevant definitions:

(i) $\{2n-51\}$. (ii) $\{n^2-n\}$. (iii) $\{1/n+1/n^2\}$. (iv) $\{1/\sqrt{n}\}$.

3. State the values of lim x^n_n for all $x > -1$ (use Theorem 3.2.4). Prove that, if $x \leqslant -1$, the sequence oscillates.

3.3. Elementary theorems about sequences

Sequences are frequently constructed from other sequences, e.g. by multiplying the corresponding terms of two sequences. In this section we examine the relationship between the limits of sequences which are related in this and other ways.

Most of the results in this section are intuitively obvious. We begin with a result which tells us that, if every term of a sequence is multiplied by a constant k, then the limit is multiplied by k.

THEOREM 3.3.1. *If* $\{a_n\}$ *and* $\{b_n\}$ *are sequences such that* $b_n = ka_n$ *for all n, where k is some real number, and if* $\lim a_n = l$, *then* $\lim b_n = kl$.

Proof. First suppose that $k \neq 0$.

Given $\varepsilon > 0$ also $\varepsilon/|k| > 0$, so that there is a natural number N such that $|a_n-l| < \varepsilon/|k|$ for all $n \geqslant N$, since $\lim a_n = l$. Thus for all $n \geqslant N$,

$$|b_n-kl| = |ka_n-kl| = |k(a_n-l)| = |k|\,|a_n-l| < |k|\varepsilon/|k| = \varepsilon.$$

Thus $\lim b_n = kl$.

Now suppose that $k = 0$. Then $b_n = 0$ for all n, and so $\lim b_n = 0$ by Theorem 3.2.4 (ii). Thus $\lim b_n = kl$. □

THEOREM 3.3.2. *If the sequence* $\{a_n\}$ *is convergent, then it is a bounded set.*

Proof. Suppose $\lim a_n = l$. Then, according to the definition of a limit, we may find a natural number N such that for all $n \geqslant N$, $|a_n - l| < 1$.

Thus for all $n \geqslant N$, $|a_n| \leqslant |l| + 1$.

Now the set $\{a_1, a_2, a_3, \ldots, a_{N-1}\}$ is a finite set and so it has a greatest member, say a_k. Put $M = \max(|l|+1, a_k)$, where $\max(x, y)$ means the greater of x and y. Then $|a_n| \leqslant M$ for all n. So $\{a_n\}$ is bounded. □

THEOREM 3.3.3. *$\{a_n\}$ and $\{b_n\}$ are sequences with* $\lim a_n = l$, $\lim b_n = m$

(i) *If $\{c_n\}$ is the sequence defined by $c_n = a_n + b_n$, then* $\lim c_n = l+m$.

(ii) *If $\{d_n\}$ is the sequence defined by $d_n = a_n - b_n$, then* $\lim d_n = l-m$.

(iii) *If $\{e_n\}$ is the sequence defined by $e_n = a_n b_n$, then* $\lim e_n = lm$.

(iv) *If $\{f_n\}$ is the sequence defined by $f_n = 1/a_n$, then* $\lim f_n = 1/l$ provided that $l \neq 0$.

Proof. (i) Given $\varepsilon > 0$, there is a natural number N with $|a_n - l| < \varepsilon/2$ for all $n \geqslant N$, and a natural number M with $|b_n - m| < \varepsilon/2$ for all $n \geqslant M$. Put $P = \max(N, M)$. Then, using Theorem 3.2.1, for all $n \geqslant P$,

$$\begin{aligned} |c_n - (l+m)| &= |(a_n - l) + (b_n - m)| \\ &\leqslant |a_n - l| + |b_n - m| < \varepsilon/2 + \varepsilon/2 = \varepsilon. \end{aligned}$$

Thus $\lim c_n = l+m$.

The proof of (ii) is similar and is left as an exercise.

(iii) Since $\{a_n\}$ is a convergent sequence, by Theorem 3.3.2 it is bounded, so that we can find a K with $|a_n| \leqslant K$ for all n. Now

$$\begin{aligned} |e_n - lm| = |a_n b_n - lm| &= |a_n(b_n - m) + (a_n - l)\, m| \\ &\leqslant K|b_n - m| + |m|\,|a_n - l|. \end{aligned}$$

Given $\varepsilon > 0$, there is a natural number N with $|a_n - l| < \varepsilon/(2|m|+1)$ for all $n \geqslant N$, and a natural number M with $|b_n - m| < \varepsilon/2K$ for all

$n \geqslant M$. Put $P = \max(M, N)$. Then for all $n \geqslant P$,

$$|e_n - lm| \leqslant K|b_n - m| + |m||a - l| < K\frac{\varepsilon}{2K} + |m|\frac{\varepsilon}{2|m|+1} < \varepsilon.$$

Therefore $\lim e_n = lm$.

(iv) Since $\lim a_n = l$ and $l \neq 0$, there is a natural number N with $|a_n - l| < |l|/2$ for all $n \geqslant N$. Now $|l - a_n| + |a_n| \geqslant |l - a_n + a_n| = |l|$ so that $|a_n| > |l| - |l|/2 = |l|/2$ for all $n \geqslant N$. Given $\varepsilon > 0$, there is a natural number M with $|a_n - l| < \varepsilon l^2/2$ for all $n \geqslant M$.

Put $P = \max(N, M)$. Then for all $n \geqslant P$,

$$|f_n - 1/l| = \left|\frac{l - a_n}{la_n}\right| < \left|\frac{\varepsilon l^2}{2|l||a_n|}\right| < \frac{2\varepsilon l^2}{2l^2} = \varepsilon.$$

Therefore $\lim f_n = 1/l$. □

COROLLARY. *$\{a_n\}$ and $\{b_n\}$ are sequences with* $\lim a_n = l$, $\lim b_n = m$. *If $\{g_n\}$ is the sequence defined by $g_n = a_n/b_n$, then* $\lim g_n = l/m$ *provided that* $m \neq 0$.

Proof. We write $g_n = a_n \cdot 1/b_n$ and use parts (iii) and (iv) of the Theorem. □

THEOREM 3.3.4. *If $\{a_n\}$ and $\{b_n\}$ are sequences such that* $\lim a_n = 0$ *and $\{b_n\}$ is bounded, and if the sequence $\{c_n\}$ is defined by $c_n = a_n b_n$, then* $\lim c_n = 0$.

Proof. Since $\{b_n\}$ is bounded, there is a number K with $|b_n| \leqslant K$ for all n. Given $\varepsilon > 0$, there is a natural number N such that for all $n \geqslant N$, $|a_n| < \varepsilon/K$. Then for all $n \geqslant N$,

$$|c_n - 0| = |a_n b_n| = |a_n|\, b_n < (\varepsilon/K) \cdot K = \varepsilon.$$

Therefore $\lim c_n = 0$. □

THEOREM 3.3.5. *$\{a_n\}$ is a sequence with* $\lim a_n = \infty$. *$\{b_n\}$ is the sequence defined by $b_n = 1/a_n$. Then* $\lim b_n = 0$.

Proof. Given $\varepsilon > 0$, since $\lim a_n = \infty$, we can find a natural number N, such that for all $n \geqslant N$, $a_n > 1/\varepsilon$. Then for all $n \geqslant N$, $|b_n - 0| = 1/a_n < \varepsilon$. Therefore $\lim b_n = 0$. □

We note that the converse of this theorem is not true; that is if $\lim a_n = 0$ and if $b_n = 1/a_n$ we may not conclude that $\lim b_n = \infty$. For if $a_n = (-1)^n/n$ then $\lim a_n = 0$. But $b_n = (-1)^n n$ and so $\{b_n\}$ oscillates.

THEOREM 3.3.6. *$\{a_n\}$ is a sequence and* $\lim a_n = l$.

(i) *If, for all n, $a_n \leqslant M$ then $l \leqslant M$.*

(ii) *If, for all n, $a_n \geqslant M$ then $l \geqslant M$.*

Proof. (i) Suppose $l > M$; then $l - M > 0$. Since $\lim a_n = l$, there is a natural number N such that for all $n \geqslant N$, $|a_n - l| < l - M$. Thus for all $n \geqslant N$, $a_n > M$. This contradicts the conditions of the theorem.

The proof of (ii) is similar and is left as an exercise. □

COROLLARY. *$\{a_n\}$ and $\{b_n\}$ are sequences, and* $\lim a_n = l$, $\lim b_n = m$. *If for all n $a_n \leqslant b_m$, then $l \leqslant m$.*

Proof. Define the sequence $\{c_n\}$ by $c_n = b_n - a_n$. Then $c \geqslant 0$ for all n. Therefore, using the Theorem, $\lim c_n \geqslant 0$. Hence, by Theorem 3.3.3, $\lim b_n - \lim a_n = \lim c_n \geqslant 0$. Thus $\lim a_n \geqslant \lim b_n$, or $l \geqslant m$. □

THEOREM 3.3.7. (i) *$\{a_n\}$ and $\{b_n\}$ are sequences, and* $\lim a_n = \infty$. *If $b_n \geqslant a_n$ for all n, then* $\lim b_n = \infty$.

(ii) *$\{a_n\}$, $\{b_n\}$ and $\{c_n\}$ are sequences with* $\lim a_n = l$, $\lim c_n = \infty$. *If $c_n = a_n b_n$ for all values of n, then* $\lim b_n = \infty$ *provided that $l > 0$.*

Proof. (i) Given any K, there is a natural number N, such that for all $n \geqslant N$, $a_n > K$. Thus for all $n \geqslant N$, $b_n \geqslant a_n > K$. Thus $\lim b_n = \infty$.

(ii) Since $l > 0$, there is a natural number N such that for all $n \geqslant N$, $|a_n - l| < l$ and so $a_n > 0$. By Theorem 3.3.3 $\lim 1/a_n = 1/l$.

Therefore, by Theorem 3.3.2, $\{1/a_n\}$ is bounded. Thus there is an X, with $0 < 1/a_n \leqslant X$ for all $n \geqslant N$. Given any $K > 0$, there is a natural number M such that for all $n \geqslant M$, $c_n > KX$. Put $p = \max(M, N)$. Then for all $n \geqslant p$, $b_n = c_n/a_n > KX/X = K$. Therefore $\lim b_n = \infty$. □

THEOREM 3.3.8. *$\{a_n\}$, $\{b_n\}$ and $\{c_n\}$ are sequences and* $\lim a_n = \lim c_n = l$. *If for all n, $a_n \leqslant b_n \leqslant c_n$, then* $\lim b_n = l$.

Proof. Given $\varepsilon > 0$, there is a natural number N such that, for all $n \geqslant N$, $a_n > l-\varepsilon$, and there is a natural number M such that, for all $n \geqslant M$, $c_n < l+\varepsilon$. Put $P = \max(M, N)$. Then for all $n \geqslant P$, $l-\varepsilon < b_n < l+\varepsilon$. Therefore $\lim b_n = l$. □

THEOREM 3.3.9. *Changing a finite number of terms of a sequence does not affect the limit of the sequence.*

Proof. Since all the definitions of a limit only demand that the terms of the sequence obey certain conditions for $n \geqslant N$, we may evidently choose N large enough so that for $n \geqslant N$, none of the terms of the sequence have been changed. □

In view of this theorem, we may relax the conditions of many of the above theorems. Wherever we have demanded that some condition should hold "for all n", we may now substitute "for all sufficiently large n".

Exercises 3.3

1. Write down the nth term of each of the following sequences. If the sequence tends to a limit state its limit.

(i) $1, 1/4, 1/9, 1/16, 1/25, 1/36, \ldots,$

(ii) $-5, -5, -5, -5, -5, -5, -5, \ldots,$

(iii) $-2, 2, -2, 2, -2, 2, -2, 2, -2, \ldots,$

(iv) $3, 3\frac{1}{2}, 3\frac{3}{4}, 3\frac{7}{8}, 3\frac{15}{16}, \ldots,$

(v) $\sqrt{1}, \sqrt{2}, \sqrt{3}, \sqrt{4}, \sqrt{5}, \sqrt{6}, \ldots,$

(vi) $1/\sqrt{1}, 1/\sqrt{3}, 1/\sqrt{5}, 1/\sqrt{7}, 1/\sqrt{9} \ldots.$

2. Find the limits of the following sequences:

(i) $\left\{\dfrac{n+1}{n^2-3}\right\}$, [*Hint:* divide numerator and denominator by n^2].

(ii) $\left\{\dfrac{2n^3+5}{n^2+1}\right\}$

(iii) $\left\{\dfrac{3n^2-6n+5}{4-3n+7n^2}\right\}$.

3. Prove Theorem 3.3.3 (ii).

4. Prove Theorem 3.3.6 (ii).

5. $\{a_n\}$ is a sequence with $\lim a_n = -\infty$. If $\{b_n\}$ is the sequence defined by $b_n = 1/a_n$, show that $\lim b_n = 0$.

6. $\{a_n\}$ is a sequence with $\lim a_n = l$. If $a_n < M$ for all n, then $l < M$.
Compare this statement with Theorem 3.3.6.(i). Is this statement true? Give a reason for your answer.

7. $\{a_n\}$ is a sequence with $\lim a_n = \infty$. If $\{b_n\}$ is the sequence defined by $b_n = -a_n$ for all n, show that $\lim b_n = -\infty$.

8. Give examples of convergent sequences for which:

(i) the limit is different from all the terms of the sequence;

(ii) the limit is equal to all the terms of the sequence;

(iii) the limit is equal to an infinite number of terms of the sequence, but not all of them;

(iv) the limit is equal to a finite number of terms of the sequence.

9. $\{a_n\}$ and $\{b_n\}$ are sequences and $\lim a_n = l$, $\lim b_n = \infty$. If $\{c_n\}$ is the sequence defined by $c_n = a_n+b_n$, prove that $\lim c_n = \infty$. (Use Theorem 3.3.2.)

10. $\{a_n\}$ and $\{b_n\}$ are sequences, and $\lim a_n = \infty$, $\lim b_n = -\infty$. $\{c_n\}$ is the sequence defined by $c_n = a_n+b_n$ for all n. Give examples to show that the following are all possible: (i) $\lim c_n = \infty$; (ii) $\lim c_n = -\infty$, (iii) $\{c_n\}$ is convergent; (iv) $\{c_n\}$ oscillates.

11. $\{a_n\}$ and $\{b_n\}$ are sequences. $b_n \neq 0$ for all n and $\lim a_n = \lim b_n = 0$. $\{c_n\}$ is the sequence defined by $c_u = a_n/b_n$ for all n. Give examples to show that the following are all possible: (i) $\lim c_n = \infty$; (ii) $\lim c_n = -\infty$; (iii) $\{c_n\}$ is convergent; (iv) $\{c_n\}$ oscillates.

12. $\{a_n\}$, $\{b_n\}$, and $\{c_n\}$ are sequences such that for all n, $c_n = a_nb_n$.

(i) If neither $\{a_n\}$ nor $\{b_n\}$ is convergent, can $\{c_n\}$ be convergent?

(ii) If only one of $\{a_n\}$ and $\{b_n\}$ is convergent, can $\{c_n\}$ be convergent?

13. Which of the following statements are true:

(i) If $\{a_n\}$ is an increasing sequence, then $\lim a_n = \infty$.

(ii) If $\{a_n\}$ is not an increasing sequence, then $\{a_n\}$ does not tend to infinity.

(iii) If $\lim a_n = \infty$, then $\{a_n\}$ is not bounded.

Justify your answers.

3.4. Behaviour of monotonic sequences

Monotonic sequences are of great importance because it is relatively easy to test whether they are convergent or not. The following theorems indicate what is necessary to prove the convergence of a monotonic sequence, and are of great importance.

THEOREM 3.4.1. *If* $\{a_n\}$ *is an increasing sequence, then* $\{a_n\}$ *is convergent if it is bounded above. Otherwise* $\lim a_n = \infty$.

Proof. First suppose $\{a_n\}$ is not bounded above. Then, given any real number K, we can find a natural number N with $a_N > K$. Since $\{a_n\}$ is increasing, for all $n \geqslant N$, $a_n \geqslant a_N > K$. Therefore $\lim a_n = \infty$.

Now suppose $\{a_n\}$ is bounded above. Then, by Theorem 2.4.2, we may find a least upper bound which we shall call l. Clearly $l \geqslant a_n$ for all n.

Given $\varepsilon > 0$, since l is the *least* upper bound, there is a natural number N such that $l - a_N < \varepsilon$. Since $\{a_n\}$ is increasing, for all $n \geqslant N$, $l - \varepsilon < a_N \leqslant a_n \leqslant l$, so that $|a_n - l| < \varepsilon$.

Thus $\lim a_n = l$. □

THEOREM 3.4.2. *If* $\{a_n\}$ *is a decreasing sequence, then* $\{a_n\}$ *is convergent if it is bounded below. Otherwise* $\lim a_n = -\infty$.

Proof. Define the sequence $\{b_n\}$ by $b_n = -a_n$. Then $\{b_n\}$ is an increasing sequence.

First suppose $\{a_n\}$ is not bounded below. Then $\{b_n\}$ is not bounded above. Therefore, by Theorem 3.4.1, $\lim b_n = \infty$.

Thus $\lim a_n = -\infty$ (see exercises 3.3, question 7).

Now suppose $\{a_n\}$ is bounded below. Then $\{b_n\}$ is bounded above. Therefore, by Theorem 3.4.1, $\{b_n\}$ is convergent. Thus, by Theorem 3.3.1, $\{a_n\}$ is convergent. □

We note that the above theorems apply to monotonic sequences only. Other sequences may be bounded without being convergent.

Exercises 3.4

1. Give an example of (i) an increasing sequence which tends to zero, (ii) a decreasing sequence which tends to 4.

2. Prove that an increasing sequence is always bounded below.

3. Give an example of a non-increasing sequence which is not bounded above but which does not tend to infinity.

4. Give an example of a bounded sequence which is neither increasing nor decreasing, and which (i) is convergent, (ii) oscillates.

5. We know that the sequence $\{a_n\}$ is convergent and that its limit is equal to its upper bound. Can we conclude that $\{a_n\}$ is an increasing sequence?

3.5. Sequences defined by recurrence relations

Consider the sequence

$$1, 2, 5, 26, 677, 458330, \ldots .$$

The sequence is constructed in the following manner. Each member of the sequence is obtained by squaring the preceding member and adding 1. Thus $2 = 1^2+1$, $5 = 2^2+1$, $26 = 5^2+1$, and so on. It would be difficult to give a formula for the general term of this sequence. But we may define the sequence by means of the following statements:

$$a_{n+1} = a_n^2+1 \quad \text{for all } n,$$
$$a_1 = 1.$$

This is an inductive definition. It defines every term of the sequence precisely because the set of natural numbers is inductive. The first statement is called a *recurrence relation*, i.e. it tells us how to construct any term out of one (or more) preceding terms.

Suppose we are interested in $\lim a_n$. It seems clear that for this sequence $\lim a_n = \infty$. We prove that this is so to illustrate one of the methods of dealing with sequences defined by recurrence relations.

$a_1 = 1$. Since $a_{n+1} = a_n^2+1$, $a_n \geqslant 1$ for all n. We shall prove by induction that $a_n \geqslant n$ for all n. Suppose $a_n \geqslant n$. Then $a_{n+1} = a_n^2+1 \geqslant n^2+1 \geqslant n+1$. Also $a_1 = 1$. Therefore, by mathematical induction $a_n \geqslant n$ for all n. Thus, by Theorem 3.3.7, $\lim a_n = \infty$.

The Fibonacci sequence

$$1, 1, 2, 3, 5, 8, 13, 21, 34, \ldots,$$

is a well known sequence defined by means of a recurrence relation. For this sequence we have

$$a_{n+2} = a_{n+1}+a_n \quad \text{for all } n.$$
$$a_1 = 1, \quad a_2 = 1.$$

We show again that $\lim a_n = \infty$. We first prove by induction that $a_n \geqslant n$ for all $n \geqslant 5$. We have $a_3 = 1+1 = 2$, $a_4 = 1+2 = 3$, $a_5 = 2+3 = 5$, $a_6 = 3+5 = 8$. Suppose that $a_n \geqslant n$ for all n satisfying $5 \leqslant n \leqslant N$. Then, if $N \geqslant 6$, $a_{N+1} = a_N + a_{N-1} \geqslant N+(N-1) > N$. Also $a_5 = 5$, $a_6 = 8$. Therefore, by mathematical induction, $a_n \geqslant n$ for all $n \geqslant 5$. Thus, by Theorem 3.3.7, modified in view of Theorem 3.3.9, $\lim a_n = \infty$.

We now consider the sequence defined by

$$19a_{n+1} = a_n^3+30.$$

This sequence begins

$$1, 1\tfrac{12}{19}, 1\tfrac{105\,240}{130\,321}, \ldots .$$

We shall prove that this sequence is convergent and find its limit.

We first prove by induction that $a_n < 2$ for all n. Suppose that $a_n < 2$. Then $19a_{n+1} = a_n^3+30 < 2^3+30 = 38$. Thus $a_{n+1} < 2$. Also $a_1 = 1 < 2$. Therefore, by induction, $a_n < 2$ for all n. Also $a_n \geqslant 1$.

We now prove that the sequence is increasing.

$$19(a_{n+1}-a_n) = a_n^3+30-19a_n = (a_n-2)(a_n-3)(a_n+5).$$

But $1 \leqslant a_n < 2$. Therefore $19(a_{n+1}-a_n) > 0$ for all n, and $\{a_n\}$ is an increasing sequence. Thus we have an increasing sequence which is bounded above by 2, so that, according to Theorem 3.4.1, $\{a_n\}$ is convergent.

Define the sequence $\{b_n\}$ by $b_n = a_{n+1}$. Then $\lim a_n = \lim b_n$, and so, by Theorem 3.3.3, $\lim(a_n-b_n) = 0$. Therefore

$$\lim\,(a_n-2)(a_n-3)(a_n+5) = 0,$$

so that again by Theorem 3.3.3, $(l-2)(l-3)(l+5) = 0$, where $l = \lim a_n$. But in view of Theorem 3.3.6, $1 \leqslant l \leqslant 2$. Thus $\lim a_n = 2$.

Iterative processes and the solutions of equations. Convergent sequences defined by recurrence relations have important practical applications, especially when the limit is unknown.

For example, suppose we have to find a root of the cubic equation

$$2x^3-16x+15 = 0.$$

We may define a sequence by the recurrence relation

$$16a_{n+1} = 2a_n^3+15$$

$$a_1 = 1.$$

Then it may be shown that $\{a_n\}$ is an increasing sequence which is bounded above. As in the previous example, we may show that the limit of this sequence is a solution of the cubic equation.

Thus if we evaluate the terms of $\{a_n\}$ one by one, using the recurrence relation, we obtain ever better approximations to one of the roots of the equation. Such a method of approximation is called an iterative process.

For the completion of this example in detail, see question 1 of the exercises for Chapter 6.

Question 4 of exercises 3.5 below gives an example of an iterative process for the evaluation of square roots.

Exercises 3.5

1. If the sequence $\{a_n\}$ is defined by the recurrence relation

$$19a_{n+1} = a_n^3+30$$

find the limit of the sequence in each of the following cases: (i) $a_1 = 2\frac{1}{2}$; (ii) $a_1 = 0$; (iii) $a_1 = 5$; (iv) $a_1 = -7$.

2. The sequence $\{b_n\}$ is defined by the recurrence relation

$$7b_{n+1} = b_n^3+6$$

and a given value of b_1. If $b_1 = 1\frac{1}{2}$,show that for all n, $b_{n+1} < b_n$, and also that $\{b_n\}$ is bounded below. Hence show that $\{b_n\}$ is convergent, and find $\lim b_n$.

Investigate similarly when $b_1 = 3$.

3. The Fibonacci sequence is defined by the recurrence relation

$$a_{n+2} = a_{n+1}+a_n$$

with $a_1 = a_2 = 1$. Verify that the nth term of this sequence is given by the formula

$$a_n = \frac{1}{\sqrt{5}}\left[\left(\frac{1+\sqrt{5}}{2}\right)^n-\left(\frac{1-\sqrt{5}}{2}\right)^n\right]$$

Hence evaluate $\lim\,(a_{n+1}/a_n)$

4. Show that $(a+k/a)^2 \geqslant 4k$. Hence show that, if $k>0$ and if $a_1 > \sqrt{k}$, then the sequence $\{a_n\}$ defined by the recurrence relation

$$a_{n+1} = \tfrac{1}{2}\left(a_n+\frac{k}{a_n}\right)$$

is bounded below by $\sqrt{k}$. Deduce that $\{a_n\}$ is a decreasing sequence. Hence show that the sequence is an iterative process for evaluating $\sqrt{k}$. Use it to evaluate $\sqrt{5}$ to two places of decimals.

3.6. More sequences and their limits

THEOREM 3.6.1. *If k is a constant greater than zero, then the sequence $\{a_n\}$ defined by $a_n = k^{1/n}$ is convergent, and* $\lim a_n = 1$.

Proof. First suppose $k > 1$. The sequence is decreasing and is bounded below by 1, so it is convergent. Since $a_n > 1$ for all n, we may write $a_n = 1+x_n$, where $x_n > 0$ for all n. Then $k = (1+x_n)^n \geqslant 1+nx_n$ by the binomial theorem, since $x_n > 0$. Thus $0 < x_n \leqslant (k-1)/n$. But $\lim (k-1)/n = 0$. Therefore, by Theorem 3.3.8, $\lim x_n = 0$. Thus, by Theorem 3.3.3, $\lim a_n = 1$.

Now suppose $k < 1$. Then $1/k > 1$. Thus, by the first parts, $\lim [(1/k)^{1/n}] = 1$. Therefore, by Theorem 3.3.3, $\lim a_n = 1$.

If $k = 1$, the theorem is trivial. □

THEOREM 3.6.2. *If* $a_n = n^{1/n}$, *the sequence* $\{a_n\}$ *is convergent, and* $\lim a_n = 1$.

Proof. $a_n > 1$ for all n. So put $a_n = 1+x_n$, where $x_n > 0$ for all n. Then.

$$n = (1+x_n)^n \geqslant 1+nx_n+\tfrac{1}{2}n(n-1)x_n^2,$$

by the binomial theorem, for all $n \geqslant 2$ since $x_u \geqslant 0$. Hence $n-1 > \frac{1}{2}n(n-1)\, x_n^2$. Thus $0 < x_n^2 < 2/n$ and $0 < x_n^2 < \sqrt{(2/n)}$. But $\lim \sqrt{(2/n)} = 0$. Thus, by Theorem 3.3.8, $\lim x_n = 0$. Therefore $\lim a_n = 1$. □

We advise the reader to note Theorem 3.6.2 particularly, since the result is perhaps difficult to forecast.

THEOREM 3.6.3. *If* $\{a_n\}$ *is the sequence defined by* $a_n = x^n/n^c$, *where* $x > 1$ *and c is a positive rational, then* $\lim a_n = \infty$.

Proof. Put $x = 1+z$ so that $z > 0$. Choose a natural number p so that $p > c$. By the binomial theorem, since $z > 0$,

$$(1+z)^n > \frac{n(n-1)\ldots(n-p+1)}{p!} z^p \quad \text{for all} \quad n \geqslant p.$$

Thus $$\frac{(1+z)^n}{n^c} > \frac{1}{p!}\left(1-\frac{1}{n}\right)\left(1-\frac{2}{n}\right)\ldots\left(1-\frac{p-1}{n}\right)n^{c-p}.$$

Therefore $$b_n = \frac{p!}{\left(1-\frac{1}{n}\right)\left(1-\frac{2}{n}\right)\ldots\left(1-\frac{p-1}{n}\right)} a_n \geqslant n^{p-c}$$

$$\text{for all} \quad n \geqslant p.$$

But lim $n^{p-c} = \infty$. Therefore, by Theorem 3.3.7(i) and 3.3.9, lim $b_n = \infty$. But

$$\lim \left[\frac{p!}{\left(1-\frac{1}{n}\right)\cdots\left(1-\frac{p-1}{n}\right)} \right] = p!$$

Therefore, by Theorem 3.3.7(ii), lim $a_n = \infty$. □

The sequences $\{x^n\}$ and $\{n^c\}$ both tend to infinity. So we express the result of this theorem informally by saying that the sequence $\{x^n\}$ tends to infinity faster than the sequence $\{n\}^c$.

THEOREM 3.6.4. *If $\{a_n\}$ is the sequence defined by $a_n = x^n/n!$, and if $x > 1$, then* lim $a_n = 0$.

Proof. For all $n \geq x^2$,

$$a_{2n} = \frac{(x^2)^n}{2n\cdot 2n-1\ldots n+1\cdot n!} = \left(\frac{x^2}{2n}\right)\left(\frac{x^2}{2n-1}\right)\cdots\left(\frac{x^2}{n+1}\right)\frac{1}{n!} < \frac{1}{n!}$$

and

$$a_{2n+1} = \frac{x}{2n+1}\, a_{2n} < \frac{1}{n!}.$$

Define the sequence $\{b_n\}$ by $b_{2n} = b_{2n+1} = 1/n!$. Then for all $n \geq 2x^2$, $b_n > a_n$. But lim $b_n = 0$. Therefore, by Theorems 3.3.6 (corollary) and 3.3.9, lim $a_n = 0$.

Once more the sequences $\{x^n\}$ and $\{n!\}$ both tend to infinity. So we express the result of this theorem informally by saying that the sequence $\{n!\}$ tends to infinity faster than the sequence $\{x^n\}$.

Sequences involving trigonometric functions. The reader will be familiar with the trigonometric functions sine, cosine, tangent, and so on, and with many of their elementary properties. For example, he will know that, for all real values of the angle x, $|\sin x| \leq 1$, $|\cos x| \leq 1$. He will also be familiar with the use of radian measure.

The problem of providing satisfactory definitions for the trigonometric functions is of interest, and is dealt with fully in Chapter 7, where many of the properties of these functions are proved.

For the time being it will be sufficient to assume a working knowledge of trigonometric functions without inquiring too deeply into the foundations of this knowledge. We shall at present be using these functions purely as examples, and not to develop the theory.

By way of example, we discuss two sequences involving sines.

$$(1) \qquad \frac{\sin x}{1}, \frac{\sin 2x}{2}, \frac{\sin 3x}{3}, \frac{\sin 4x}{4}, \frac{\sin 5x}{5}, \ldots .$$

Here the nth term of the sequence $\{a_n\}$ is $\sin nx/n$. Since for all n and all x, $|\sin nx| \leqslant 1$,

$$-1/n \leqslant a_n \leqslant 1/n \text{ for all } n.$$

Therefore, by Theorem 3.3.8, $\lim a_n = 0$.

$$(2) \qquad \sin x, \sin 2x, \sin 3x, \sin 4x, \sin 5x, \ldots .$$

Here the nth term of the sequence $\{a_n\}$ is $\sin nx$. The behaviour of this sequence depends on the value of x. If x takes one of the values 0, $\pm\pi$, $\pm 2\pi$, $\pm 3\pi$, $\pm 4\pi$, $\ldots$, then $a_n = 0$ for all n, so that $\lim a_n = 0$.

But if, for example, $x = \pi/3$, then the sequence becomes

$$\sqrt{3}/2, \sqrt{3}/2, 0, -\sqrt{3}/2, -\sqrt{3}/2, 0, \sqrt{3}/2, \sqrt{3}/2, 0, \ldots,$$

and clearly $\{a_n\}$ oscillates.

In fact it may be shown that, if x takes any value other than $m\pi$, the sequence will oscillate.

Exercises 3.6

1. Discuss the behavior of $\{a_n\}$ as n tends infinity in the following cases:

(i) $a_n = \dfrac{123^n}{n!}$. (ii) $a_n = \dfrac{n^{58}}{123^n}$. (iii) $a_n = \dfrac{n^{58}}{n!}$.

(iv) $a_n = (n^2)^{1/n}$. (v) $a_n = \dfrac{1}{(-3)^n n^2}$. (vi) $a_n = \dfrac{\sin(n^2)}{\sqrt{n}}$.

(vii) $a_n = \tan n$. (viii) $a_n = \dfrac{3^{n!}}{n!}$.

2. The sequence $\{a_n\}$ is defined by $a_n = x^n/n^c$, where c is a positive rational. What is the behaviour of $\{a_n\}$ as n tends to infinity: (i) If $|x| < 1$? (ii) if $x = 1$? (iii) if $x < -1$?

3. If $\{a_n\}$ and $\{b_n\}$ are sequences which tend to infinity and $\lim [a_n/b_n] = 0$, we say that $\{b_n\}$ tends to infinity faster than $\{a_n\}$. Prove that, if $\{b_n\}$ tends to infinity faster than $\{a_n\}$ and if $\{c_n\}$ tends to infinity faster than $\{b_n\}$, then $\{c_n\}$ tends to infinity faster than $\{a_n\}$.

3.7. Upper and lower limits

In this section we shall first consider ways of describing the behaviour as n tends to infinity of oscillating sequences. We shall then prove an important test for the convergence of a sequence, the general principle of convergence.

Oscillating sequences. Consider the sequence

$$2\tfrac{1}{2},\ -1\tfrac{1}{2},\ 2\tfrac{1}{4},\ -1\tfrac{1}{4},\ 2\tfrac{1}{8},\ -1\tfrac{1}{8},\ 2\tfrac{1}{16},\ -1\tfrac{1}{16},\ \ldots .$$

This sequence is bounded and oscillating. Its least upper bound is $2\frac{1}{2}$, but $2\frac{1}{2}$ is in no way suitable for describing the behaviour of the sequence as n tends to infinity. $2\frac{1}{2}$ is almost the least upper bound by accident; if we omitted the first term we would not materially affect the nature of the sequence but we *would* change the least upper bound. Consequently we seek other numbers which more adequately describe the behaviour of the sequence.

If we look at the larger members of the sequence we see that they approximate to 2, so that 2 is the "limit" of the larger members. Similarly, the "limit" of the smaller members is -1.

It is these ideas which we now wish to make precise.

We have already remarked that if we omit the first term of the sequence the upper bound changes. In fact it falls from $2\frac{1}{2}$ to $2\frac{1}{4}$. If we continue to omit more and more terms from the beginning of the sequence, the upper bound falls more and more and in fact forms a sequence which tends to 2. So, too, the lower bound of the sequence increases as we omit more and more terms and forms a sequence which tends to -1. We shall call 2 the upper limit of the sequence and -1 the lower limit of the sequence.

DEFINITION. *If $\{a_n\}$ is a sequence which is bounded above, we define the* nth least upper bound *of the sequence, which we shall write K_n, to be least upper bound of the set $\{a_n, a_{n+1}, a_{n+2}, \ldots\}$, i.e. the least upper bound of the set obtained by omitting the first $n-1$ terms of the sequence.*

THEOREM 3.7.1. *The sequence $K_1, K_2, K_3, K_4, \ldots,$ is decreasing.*

Proof. This theorem is more or less self-evident (see question 8 of exercises 2.4). □

Since $K_1, K_2, K_3, K_4, \ldots$ is decreasing, then, according to Theorem 3.4.2, either $\{K_n\}$ is convergent or $\lim K_n = -\infty$. Consequently we make the following definition.

DEFINITION. *If $\{a_n\}$ is a sequence which is bounded above, then* $\lim K_n$ *is called the* upper limit *of the sequence $\{a_n\}$, written $\overline{\lim}\, a_n$. If $\{a_n\}$ is not bounded above, then we define $\overline{\lim}\, a_n = \infty$.*

DEFINITION. *If $\{a_n\}$ is a sequence which is bounded below, we define the* nth greatest lower bound *of the sequence, which we shall write k_n, to be the greatest lower bound of the set $\{a_n, a_{n+1}, a_{n+2}, \ldots\}$.*

THEOREM 3.7.2. *The sequence $k_1, k_2, k_3, k_4, \ldots,$ is increasing.*

Consequently either $\{k_n\}$ is convergent or $\lim k_n = \infty$. So we make the following definition.

DEFINITION. *If $\{a_n\}$ is a sequence which is bounded below, then* $\lim k_n$ *is called the* lower limit *of the sequence $\{a_n\}$, written $\underline{\lim}\, a_n$. If $\{a_n\}$ is not bounded below, then we define $\underline{\lim}\, a_n = -\infty$.*

We note that the definitions define upper and lower limits for *all* sequences.

Upper and lower limits of convergent sequences. The following theorem is of a technical nature.

THEOREM 3.7.3. *$\{a_n\}$ is a bounded sequence and $\overline{\lim}\, a_n = \Lambda$, $\underline{\lim}\, a_n = \lambda$. Then given $\varepsilon > 0$, we can find a natural number N such that:*

(i) *$a_n < \Lambda + \varepsilon$ for all $n \geqslant N$;*

(ii) *$a_n > \Lambda - \varepsilon$ for some n as large as we please;*

(iii) *$a_n < \lambda + \varepsilon$ for some n as large as we please;*

(iv) *$a_n > \lambda - \varepsilon$ for all $n \geqslant N$.*

Proof. Define the sequences $\{K_n\}$ and $\{k_n\}$ as before. Lim $K_n = \Lambda$, so we can find a natural number M such that, $|K_M - \Lambda| < \varepsilon$. Then, for all $n \geqslant M$,

$$a_n \leqslant K_M < \Lambda + \varepsilon.$$

Since $\{K_n\}$ is a decreasing sequence, $K_m \geqslant \Lambda$ for all m. Since K_m is a *least* upper bound, there is an $n \geqslant m$ for which $a_n > K_m - \varepsilon$. Therefore $a_n > \Lambda - \varepsilon$. Since $\{k_n\}$ is an increasing sequence, $k_m \leqslant \lambda$ for all m. Since k_m is a *greatest* lower bound, there is an $n \geqslant m$ for which $a_n < k_m + \varepsilon$. Therefore $a_n < \lambda + \varepsilon$.

Finally, lim $k_n = \lambda$, so we can find a natural number P such that $|k_P - \lambda| < \varepsilon$. Then, for all $n \geqslant P$, $a_n \geqslant k_P > \lambda - \varepsilon$. Put $N = \max(M, P)$ and the result follows. □

THEOREM 3.7.4. *A bounded sequence $\{a_n\}$ is convergent if and only if its upper and lower limits are equal. If it is convergent, then* $\lim a_n = \overline{\lim}\, a_n = \underline{\lim}\, a_n$.

Proof. Let $\overline{\lim}\, a_n = \Lambda$, $\underline{\lim}\, a_n = \lambda$. First, suppose that $\{a_n\}$ is convergent and let $\lim a_n = l$. Given $\varepsilon > 0$, there is a natural number N such that $|a_n - l| < \varepsilon$ and for which $a_n < \Lambda + \varepsilon$, for all $n \geqslant N$.

By Theorem 3.7.3, for some $n \geqslant N$, say $n = p$, $a_n > \Lambda - \varepsilon$. Thus

$$|\Lambda - l| \leqslant |\Lambda - a_p| + |a_p - l| < \varepsilon + \varepsilon = 2\varepsilon.$$

Now Λ and l are both fixed numbers, independent of the choice of ε, and the difference between them is less than 2ε, which may be made as small as we please. Therefore $\Lambda = l$. Similarly, $\lambda = l$. Thus $\Lambda = \lambda = l$.

Now suppose $\Lambda = \lambda$. By Theorem 3.7.3 we can find a natural number N with

$$a_n < \Lambda+\varepsilon \text{ for all } n \geqslant N$$

and

$$a_n > \Lambda-\varepsilon \text{ for all } n \geqslant N.$$

Thus for all $n \geqslant N$, $|a_n-\Lambda| < \varepsilon$, and so $\lim a_n = \Lambda$. □

(The reader is advised to make sure he understands why the following argument is false. $\lim b_n = l$. Therefore there is a natural number N with $|b_n-l| < \varepsilon$ for all $n \geqslant N$. But ε is as small as we please, and so $b_n = l$.)

The general principle of convergence. If we have a sequence, and it is obvious by inspection what the limit of this sequence is, then we may prove the sequence is convergent by applying the definition of a limit. However, if the limit is not obvious, it is necessary to discover in some other way whether the sequence is convergent. In section 3.4 we gave tests for the convergence of such a sequence. Another such test is the general principle of convergence, and it may be applied when the sequence is not monotonic.

DEFINITION. *A sequence $\{a_n\}$ is called a* Cauchy sequence *if, given $\varepsilon > 0$, we can always find a natural number N such that*

$$|a_n-a_m| < \varepsilon \quad \textit{for all} \quad n \geqslant N, m \geqslant N.$$

THEOREM 3.7.5. *A sequence is convergent if and only if it is a Cauchy sequence.*

Proof. First suppose the sequence is convergent. Call the sequence $\{a_n\}$. Then, given $\varepsilon > 0$, we can find a natural number N such that for all $n \geqslant N$, $|a_n - l| < \varepsilon/2$, where l is the limit of the sequence. Thus if $n \geqslant N$ and $m \geqslant N$,

$$|a_n - a_m| \leqslant |a_n - l| + |l - a_m| < \varepsilon/2 + \varepsilon/2 = \varepsilon.$$

Therefore $\{a_n\}$ is a Cauchy sequence.

Now suppose $\{a_n\}$ is a Cauchy sequence. We first show that $\{a_n\}$ is bounded. We can find a natural number M such that, for all $n \geqslant M$, $|a_n - a_M| < 1$. Therefore the set $\{a_M, a_{M+1}, a_{M+2}, a_{M+3}, \ldots\}$ is bounded. Therefore the sequence $\{a_n\}$ is bounded. So let the upper and lower limits of $\{a_n\}$ be Λ and λ respectively. Then, given $\varepsilon > 0$, there is a natural number N such that $|a_n - a_m| < \varepsilon$ for all $n \geqslant N$ and $m \geqslant N$.

Now, by Theorem 3.7.3, there is a natural number greater than N, say p, for which $|a_p - \Lambda| < \varepsilon$, and there is a natural number greater than N, say q, for which $|a_q - \lambda| < \varepsilon$. So

$$|\Lambda - \lambda| \leqslant |\Lambda - a_p| + |a_p - a_q| + |a_q - \lambda| < 3\varepsilon.$$

Thus Λ and λ are two constants independent of our choice of ε, which differ from one another by less than 3ε, which may be made as small as we please. Thus $\Lambda = \lambda$. Therefore, by Theorem 3.7.4, $\{a_n\}$ is convergent. □

We now illustrate the use of the general principle of convergence. Consider the sequence

$$1, \tfrac{1}{2}, \tfrac{5}{6}, \tfrac{7}{12}, \tfrac{47}{60}, \tfrac{37}{60}, \tfrac{319}{420}, \ldots,$$

whose nth term a_n is defined by

$$a_1 = 1, \quad a_{n+1} = a_n + \frac{(-1)^n}{n+1}.$$

If $n > m$, we have

$$a_n - a_m = (-1)^m \left(\frac{1}{m+1} - \frac{1}{m+2} + \frac{1}{m+3} + \ldots + \frac{(-1)^{n-m-1}}{n}\right).$$

Now $$\left(\frac{1}{m+1}-\frac{1}{m+2}\right)+\left(\frac{1}{m+3}-\frac{1}{m+4}\right)+\ldots+\frac{(-1)^{n-m-1}}{n} > 0.$$

Also $$\frac{1}{m+1}-\left(\frac{1}{m+2}-\frac{1}{m+3}\right)-\left(\frac{1}{m+4}-\frac{1}{m+5}\right)-\ldots$$
$$+\frac{(-1)^{n-m-1}}{n} < \frac{1}{m+1}.$$

Therefore $|a_n - a_m| < 1/m$
Given $\varepsilon > 0$, we can find a natural number N with $1/N < \varepsilon$. Then, if $n > m \geqslant N$,

$$|a_n - a_m| < \frac{1}{m+1} \leqslant \frac{1}{N+1} < \varepsilon.$$

Thus the sequence is a Cauchy sequence and therefore convergent.

Subsequences

DEFINITION. *Given a sequence* $\{a_n\}$, *a* subsequence *of* $\{a_n\}$ *is any sequence formed by striking out some of the terms of* $\{a_n\}$ *without disturbing the order of the remaining terms.*

THEOREM 3.7.6. *If* $\{a_n\}$ *is convergent, then any subsequence of* $\{a_n\}$ *is convergent and has the same limit as* $\{a_n\}$.

The proof is left as an exercise.

THEOREM 3.7.7. *If* $\{a_n\}$ *is a bounded sequence whose upper limit is* Λ *and whose lower limit is* λ, *then there is a convergent subsequence of* $\{a_n\}$ *whose limit is* Λ *and another whose limit is* λ.

Proof. Since Λ is the upper limit, by Theorem 3.7.3 there is one member of the sequence which we shall call b_1 satisfying $|b_1 - \Lambda| < 1$. So, too, there is a second member of the sequence, after b_1 in the sequence, satisfying $|b_2-\Lambda| < \frac{1}{2}$. We can similarly find a b_3, after b_2 in the sequence, satisfying $|b_3-\Lambda| < \frac{1}{4}$. In this way we construct the subsequence $\{b_n\}$ and leave the reader to show that $\lim b_n = \Lambda$.

In a similar manner we construct a sequence $\{c_n\}$ whose limit is λ. □

THEOREM 3.7.8. (i) *If a sequence is not bounded above, then we may find a subsequence which tends to infinity.*

(ii) *If a sequence is not bounded below, then we may find a subsequence which tends to minus infinity.*

The proof is again left as an exercise.

Points of accumulation. Suppose we have an infinite set of real numbers. Then we may choose from that set a sequence all of whose members are different. If this sequence is convergent, then we call its limit a *point of accumulation* of the infinite set.

For example, the set $\{1/q : q$ is a natural number$\}$ has one point of accumulation, namely zero.

The set $\{x\colon \sin x = 1/n,\ n$ is a natural number$\}$ has an infinity of points of accumulation, $m\pi$ for all integers m.

The set of real numbers has every real number as a point of accumulation.

The set of natural numbers has no point of accumulation.

The last example shows that an infinite set need have no point of accumulation. However, it is readily shown that every infinite *bounded* set has at least one point of accumulation.

THEOREM 3.7.9. *Every infinite bounded set has a point of accumulation.*

Proof. Since the set is infinite it yields a sequence, all of whose terms are different. This sequence is bounded, and so it has an upper limit, Λ say.

Then by Theorem 3.7.7, Λ is the limit of some subsequence.

Therefore Λ is a point of accumulation of the set. □

Alternative proof. We give a proof which does not depend on the theory of upper and lower limits.

Let M be the least upper bound of the set. If M does not belong to the set, then we must be able to find a sequence of points of the set whose limit is M. (Otherwise M would not be the *least* upper bond.)

If M belongs to the set, remove it. If M is still the least upper bound, then, as in the preceding paragraph, the set contains a sequence whose limit is M. If M is no longer the least upper bound, let this be M_2. Continuing in this way we either conclude the proof in the manner indicated in the previous paragraph or else we construct a sequence of upper bounds $M, M_2, M_3, M_4, \ldots$, all members of the set, which is a decreasing bounded sequence, and thus has a limit. □

This theorem may also be proved by a method analogous to that used in the proof of Theorem 5.7.2.

Exercises 3.7

1. State the upper and lower limits of the following sequences:

(i) $4, -4, 3\frac{1}{2}, -4\frac{1}{2}, 3\frac{1}{4}, -4\frac{3}{4}, 3\frac{1}{8}, -4\frac{7}{8}, \ldots,$

(ii) $1, -1, 1, -1, 1, -1, 1, -1, \ldots,$

(iii) $5, 3, 6, 4, 7, 5, 8, 6, 9, 7, \ldots,$

(iv) $1, -2, 3, -4, 5, -6, 7, -8, 9, \ldots,$

(v) $\sin \pi/6, \sin 2\pi/6, \sin 3\pi/6, \sin 4\pi/6, \sin 5\pi/6, \ldots,$

(vi) $5, 1, \sqrt{5}, 2, \sqrt[3]{5}, 3, \sqrt[4]{5}, 4, \sqrt{5}, 5, \sqrt[6]{5}, 6, \sqrt[7]{5}, 7\ldots.$

2. If a sequence tends to infinity, show carefully that its upper and lower limits are both infinity.

3. If a sequence is bounded, show that its upper limit is greater than or equal to its lower limit.

4. Give an example of a set which has four points of accumulation.

5. Give an example of a set whose points of accumulation are all the natural numbers.

6. Give an example of a bounded set whose points of accumulation are all its members.

7. Give an example of a set which has an infinite number of points of accumulation not belonging to the set.

8. Prove Theorem 3.7.6.

9. Prove Theorem 3.7.8.

10. Complete the proof of Theorem 3.7.7.

11. $\{a_n\}$ and $\{b_n\}$ are bounded sequences. If $\{c_n\}$ is the sequence defined by $c_n = a_n + b_n$ for all n show that it is not necessarily true that $\overline{\lim}\, c_n = \overline{\lim}\, a_n + \overline{\lim}\, b_n$.

12. Give an example of a sequence whose upper limit is 1 and whose lower limit is zero and which has the property that, given any number l satisfying $0 \leqslant l \leqslant 1$, there is a subsequence whose limit is l.

Exercises for Chapter 3

1. Suppose that $\{a_n\}$ is a sequence and that

$$\lim a_n = l \quad l < 0.$$

Show that at most a finite number of the terms of the sequence are greater than zero. Deduce that, if $\{b_n\}$ is the sequence defined by

$$b_n = |a_n|,$$

then $\{b_n\}$ is also convergent.

2. If $\{a_n\}$ is defined by

$$a_n = \sqrt[3]{(n+1)} - \sqrt[3]{n}.$$

show that $\lim a_n = 0$. [*Hint*: use the identity $(x-y)(x^2+xy+y^2) = x^3-y^3$.]

3. If $\{a_n\}$ is an increasing sequence which is bounded above and if $\{b_n\}$ is the sequence defined by

$$b_n = \frac{a_1+a_2+a_3+\ldots+a_n}{n},$$

show that $\{b_n\}$ is convergent.

4. Show that

$$\left(1+\frac{1}{n}\right)^n = 1+1+\frac{\left(1-\frac{1}{n}\right)}{2!}+\frac{\left(1-\frac{1}{n}\right)\left(1-\frac{2}{n}\right)}{3!}+\ldots+\frac{\left(1-\frac{1}{n}\right)\ldots\left(1-\frac{n-1}{n}\right)}{n!}$$

Hence show that:

(i) $\{(1+1/n)^n\}$ is an increasing sequence;

(ii) $(1+1/n)^n < 3$ for all n.

Deduce that $\{(1+1/n)^n\}$ is convergent.

5. If $\{a_1, a_2, a_3, \ldots, a_n\}$ is a set of fixed positive real numbers, find

$$\lim_{p \to \infty} [(a_1^p + a_2^p + \ldots + a_n^p)^{1/p}]$$

6. The terms of a sequence of numbers are $u_1, u_2, u_3, \ldots, u_n. \ldots$ Explain what is meant by the statement

$$\lim_{n \to \infty} u_n = l.$$

In each of the following cases write down the first four terms of the sequence $\{u_n\}$:

(i) $u_n = \dfrac{2n^2}{2n^2-1}$;

(ii) $u_n = 4 + \dfrac{1}{n} + 3.(-1)^n$;

(iii) u_n is the *sum* of the first n terms of the GP whose first term is 1 and whose common ratio is $\frac{1}{2}$.

In each case, state whether the sequence has or has not a limit.

For each of the cases in which the sequence has a limit, state what the limit is and find the least value of n for which u_n differs from its limit by less than 1/100.

[T.C.]

7. (i) Give a careful definition of what you mean by the statement

$$\lim_{n \to \infty} a_n = L,$$

where $\{a_n\}$ is a sequence of real numbers.

Prove from your definition that if $0 < b_n < a_n$ for all n greater than some fixed N, and if $\lim a_n = 0$, then $\lim b_n = 0$.

(ii) Prove that if $k > 1$, $\lim k^{1/n} = 1$. Prove also that, if $\{k_n\}$ is a sequence of positive numbers such that $\lim k_n = L > 1$, then $\lim k_n^{1/n} = 1$.

(iii) Use the result of part (ii) to prove that

$$\lim \left(1 + \frac{1}{n^2}\right)^n = 1.$$

(You may assume that $\lim (1+1/n)^n = e$.) [T.C.]

8. State a set of axioms for the real numbers.

Explain which of your axioms is not satisfied by the rational numbers, giving an example to illustrate your answer.

State the Cauchy general principle of convergence.

If $\{x_n\}$ is a sequence of positive rationals such that $\{x_n^2\}$ converges to a rational c, show that $\{x_n\}$ satisfies the Cauchy general principle.

Give an example of such a sequence $\{x_n\}$ which does not converge to a rational.
[B.Ed.]

9. For given positive real numbers a_1, b_1, $(a_1 < b_1)$, sequences $\{a_n\}$, $\{b_n\}$ are defined inductively by

$$a_n = \sqrt{(a_{n-1}b_{n-1})}, \quad b_n = \tfrac{1}{2}(a_{n-1}+b_{n-1}), \quad n = 2, 3, \ldots$$

(i) Show that $b_n - a_n = \frac{1}{2}(\sqrt{b_{n-1}} - \sqrt{a_{n-1}})^2$ for all n.

(ii) Show on a line the relative magnitudes of a_1, b_1, a_2, b_2.

Deduce from (i) that $\{a_n\}$ and $\{b_n\}$ are monotone and converge.

(iii) Prove that $\lim a_n = \lim b_n$. [B.Ed.]

CHAPTER 4

SERIES

4.1. Introduction

In the introductory chapter we mentioned the use of infinite series in the evaluation of important constants and the definition of certain functions. In this chapter, we develop the theory of infinite series and show how the concepts may be applied to decimal representation. We leave the other applications of series until the last two chapters.

DEFINITION. *If* $\{a_n\}$ *is a sequence, then*

$$a_1+a_2+a_3+a_4+a_5+\ldots$$

is called an infinite series. *We shall write the series* Σa_n, *or, more explicitly,* $\sum_{n=1}^{\infty} a_n$, *if we want to make clear that the first term of the series is* a_1.

As with sequences, it is sometimes convenient to call the first term of a series something other than a_1. For example, we may have

$$a_0+a_1+a_2+a_3+a_4+a_5+\ldots$$

which will be written $\sum_{n=0}^{\infty} a_n$, or

$$a_3+a_4+a_5+a_6+a_7+\ldots$$

which will be written $\sum_{n=3}^{\infty} a_n$.

Consider the series

$$9+3+1+\tfrac{1}{3}+\tfrac{1}{9}+\tfrac{1}{27}+\ldots \quad \text{or} \quad \sum_{n=1}^{\infty} 3^{3-n}.$$

Our familiarity with the addition of real numbers, of course, does not enable us to find the sum of this series because the number of its terms is infinite. Consequently, we use the concepts of sequences in order to deal with the sums of infinite series.

We obtain a sequence from the above series by adding together ever more of its terms, thus:

$$9,\ 12,\ 13,\ 13\tfrac{1}{3},\ 13\tfrac{4}{9},\ 13\tfrac{13}{27},\ \ldots.$$

This sequence is called the sequence of partial sums of the series. Since the limit of this sequence is $13\frac{1}{2}$, we shall later define the sum of the series to be $13\frac{1}{2}$.

DEFINITION. *The sum of the first n terms of a series*

$$a_1+a_2+a_3+a_4+a_5+\ldots$$

is called the nth partial sum. *It is written compactly* $\sum_{r=1}^{n} a_r$ *and is often denoted by* S_n.

For example, the series above gives

$$S_1 = 9,\ S_2 = 12,\ S_3 = 13,\ S_4 = 13\tfrac{1}{3},\ \ldots,\ S_n = 13\tfrac{1}{2}-\tfrac{1}{2}\cdot 3^{3-n}.$$

The arithmetic progression. The series

$$3+8+13+18+23+28+\ldots$$

and

$$5+2-1-4-7-10-13-\ldots$$

are both constructed according to the same kind of rule. Each term is obtained from the preceding term by adding a constant, in the first case 5 and in the second -3.

DEFINITION. *An* arithmetic progression *is a series whose nth term is* $a+(n-1)d$, *where a and d are constants. d is called the* common difference *of the series.*

The nth partial sum, S_n, of an arithmetic progression may be found as follows:

$$S_n = a+(a+d)+(a+2d)+\ldots+[a+(n-2)d]+[a+(n-1)d]$$
$$= [a+(n-1)d]+[a+(n-2)d]+\ldots+(a+2d)+(a+d)+a.$$

Therefore $$2S_n = n[2a+(n-1)d]$$
and $$S_n = (n/2)[2a+(n-1)d].$$

For example, the nth partial sum of the series

$$3+8+13+18+23+28+\ldots$$

is $n(5n+1)/2$, since here $a = 3$ and $d = 5$.

The geometric progression. For the series

$$1-2+4-8+16-32+\ldots$$
and $$6+2+\tfrac{2}{3}+\tfrac{2}{9}+\tfrac{2}{27}+\ldots$$

each term is obtained from the preceding term by multiplying by a constant, in the first case -2 and in the second $1/3$.

DEFINITION. *A* geometric progression *is a series whose nth term is* ar^{n-1}, *where a and r are constants. r is called the* common ratio *of the series.*

The nth partial sum S_n of a geometric progression may be found as follows:

If $r \neq 1$,

$$S_n = a+ar+ar^2+\ldots+ar^{n-1},$$
$$rS_n = ar+ar^2+ar^3+\ldots+ar^n.$$

Therefore $$(1-r)S_n = a-ar^n$$

and $$S_n = \frac{a(1-r^n)}{1-r}.$$

If $r = 1$, $$S_n = a+a+a\ldots+a = na.$$

For example, the nth partial sum of the series

$$1-2+4-8+16-32+\ldots$$

is $\frac{1}{3}(1-(-2)^n)$ since here $a = 1$ and $r = -2$.

The harmonic series. The series

$$1+\tfrac{1}{2}+\tfrac{1}{3}+\tfrac{1}{4}+\tfrac{1}{5}+\tfrac{1}{6}+\ldots,$$

whose nth term is $1/n$, is called the *harmonic series*. There is no explicit formula for the nth partial sum of the harmonic series.

Exercises 4.1

1. Find the twentieth partial sum of the following series:

(i) $1+4+7+10+\ldots$

(ii) $240+231+222+213+\ldots$

(iii) $1+\frac{1}{5}+\frac{1}{25}+\frac{1}{125}+\ldots$

(iv) $3+12+48+192+\ldots$

(v) $4-8+16-32+\ldots$

(vi) $1-1+1-1+1-1+1-\ldots$

2. Write down the first ten partial sums of the harmonic series.

3. Show that the eighth partial sum of the harmonic series exceeds $2\frac{1}{2}$. Deduce that the sixteenth partial sum exceeds 3, and that the thirty-second exceeds $3\frac{1}{2}$.

4. Find the sum of all the different rationals x, satisfying $0 < x < 1$ and which may be written in the form p/q, where p and q are natural numbers and $q \leqslant 20$.

4.2. Convergence of a series

Consider again the series

$$9+3+1+\tfrac{1}{3}+\tfrac{1}{9}+\tfrac{1}{27}+\ldots.$$

The sequence of partial sums for this series is

$$9, 12, 13, 13\tfrac{1}{3}, 13\tfrac{4}{9}, 13\tfrac{13}{27}, \ldots .$$

The more terms of the series we add together, the further we move along the sequence of partial sums. Therefore, to give a meaning to the sum of all the terms of the series, we use the limit of the sequence of partial sums.

DEFINITION. *Σa_n is a series whose nth partial sum will be called S_n. If $\{S_n\}$ is a convergent sequence, we say that the series Σa_n is convergent and we call the limit l of the sequence of partial sums the sum to infinity of the series. We write $\Sigma a_n = l$.*

DEFINITION. *If Σa_n is a series whose sequence of partial sums is not convergent, then we say that the series Σa_n is divergent. A divergent series does not have a sum to infinity.*

For example, the partial sums of the series given at the beginning of this section form a convergent sequence whose limit is $13\frac{1}{2}$. Thus we say that the series is convergent, and write $\sum_{n=1}^{\infty} 3^{3-n} = 13\frac{1}{2}$.

The arithmetic progression. The arithmetic progression whose nth term is $a+(n-1)\,d$ has as its nth partial sum $S_n=(n/2)[2a+(n-1)\,d]$. We leave the reader to show that, unless a and d are both zero, the sequence of partial sums is not convergent. If a and d are both zero, then every term of the series is zero. Thus:

THEOREM 4.2.1. *The only convergent arithmetic progression is the trivial progression*

$$0+0+0+0+0+0+0+0+\ldots .$$

For this reason, arithmetic progressions will not be of much interest to us.

The geometric progression. We shall call a geometric progression trivial if all its terms are zero. Otherwise it is non-trivial.

THEOREM 4.2.2. *The non-trivial geometric progression* $\sum_{n=1}^{\infty} ar^{n-1}$ *is convergent if* $|r| < 1$ *and divergent if* $|r| \geqslant 1$. *When* $|r| < 1$, *we have* $\sum_{r=1}^{\infty} ar^{n-1} = a/(1-r)$.

Proof. If $r \neq 1$, the nth partial sum of the progression is

$$S_n = \frac{a}{1-r} + \frac{ar^n}{1-r}.$$

If $|r| < 1$, then $\lim r^n = 0$. So $\{S_n\}$ is convergent and $\lim S_n = (a/1-r)$. Therefore $\sum_{n=1}^{\infty} ar^n = (a/1-r)$. If $|r| > 1$ or if $r = -1$, $\{r^n\}$ either tends to plus or minus infinity or oscillates. Thus $\{S_n\}$ is not convergent, and so $\sum_{n=1}^{\infty} ar^n$ is divergent.

If $r = 1$, the nth partial sum is $S_n = na$. Thus S_n tends to infinity, and so $\sum_{n=1}^{\infty} ar^n$ is divergent. □

This result occupies a central position in the theory of infinite series since in most of what follows we shall effectively establish whether a series converges or not by reference to the behaviour of geometric progressions.

The harmonic series. When we are given a non-trivial arithmetic progression, one look should be sufficient to convince us that it is divergent. On the other hand, when we meet the series

$$1+\tfrac{1}{2}+\tfrac{1}{3}+\tfrac{1}{4}+\tfrac{1}{5}+\tfrac{1}{6}+\tfrac{1}{7}+\dots$$

it looks at least possible that the series will converge, for the terms do get smaller and the nth term does tend to zero. However:

THEOREM 4.2.3. *The harmonic series* $\sum_{n=1}^{\infty} 1/n$ *is divergent.*

Proof. Let us call the nth partial sum S_n.

$$S_1 = 1,$$
$$S_2 = 1+\tfrac{1}{2},$$
$$S_4 = 1+\tfrac{1}{2}+\tfrac{1}{3}+\tfrac{1}{4} < 1+\tfrac{1}{2}+\tfrac{1}{2} = 1+\tfrac{2}{2},$$
$$S_8 = 1+\tfrac{1}{2}+(\tfrac{1}{3}+\tfrac{1}{4})+(\tfrac{1}{5}+\tfrac{1}{6}+\tfrac{1}{7}+\tfrac{1}{8}) < 1+\tfrac{1}{2}+\tfrac{1}{2}+\tfrac{1}{2} = 1+\tfrac{3}{2}.$$

Continuing in this way the 2^nth partial sum exceeds $1+n/2$. Thus $\{S_n\}$ is not bounded, and so, by Theorem 3.3.2, is not convergent. Hence the harmonic series is divergent. □

Since this result does seem to surprise, we elaborate on what it means for the series.

Each term of the series is smaller than the one before (and more particularly the nth term tends to zero). Therefore each time we add less; but the cumulative effect of adding these ever-decreasing terms is that eventually our sum will be as large as we please—the sum will continue increasing with no ceiling. Nevertheless, we need nearly 500 000 000 terms for the sum to reach 20, and more than 3×10^{868} terms for the sum to reach 2000. Yet, no matter what target we choose, the partial sums will reach it *eventually*.

A condition for the convergence of a series

THEOREM 4.2.4. *If* Σa_n *is a convergent series, then* $\lim a_n = 0$.

Proof. Let S_n be the nth partial sum of the series. Since the series is convergent, so is $\{S_n\}$. Let $\lim S_n = l$. Define $\{T_n\}$ by $T_n = S_{n-1}$. Then

$\lim T_n = l$. Thus, by Theorem 3.3.3, $\lim(S_n - T_n) = 0$; i.e. $\lim(S_n - S_{n-1}) = 0$; i.e. $\lim a_n = 0$. □

Given a series whose nth term does not tend to zero, we may use this theorem to deduce immediately that the series is divergent.

We note that the converse of this theorem is *false*. If the nth term of a series tends to zero, the series may yet diverge, as the harmonic series demonstrates. The reader must be careful *not* to deduce the convergence of a series from the fact that its nth term tends to zero.

We conclude this section with a straightforward theorem of a technical nature.

THEOREM 4.2.5. (i) *Σa_n is a series and Σb_n is defined by $b_n = ca_n$ for all n, where c is a constant. If Σa_n is convergent, then so is Σb_n and $\Sigma b_n = c\Sigma a_n$.*

(ii) *Σa_n and Σb_n are series and Σc_n is defined by $c_n = a_n + b_n$ for all n. If Σa_n and Σb_n are convergent, then so is Σc_n and $\Sigma c_n = \Sigma a_n + \Sigma b_n$.*

Proof. (i) Let the nth partial sums of Σa_n and Σb_n be S_n and T_n respectively. Then $T_n = cS_n$ for all n. Since $\{S_n\}$ is convergent, by Theorem 3.3.1 $\{T_n\}$ is convergent and $\lim T_n = \lim S_n$.

(ii) Let the nth partial sums of Σa_n, Σb_n, and Σc_n be S_n, T_n and U_n respectively. Since $\{S_n\}$ and $\{T_n\}$ are convergent, by Theorem 3.3.3 $\{U_n\}$ is convergent and $\lim U_n = \lim S_n + \lim T_n$. □

Exercises 4.2

1. Say whether the following arithmetic, and geometric progressions are convergent or divergent. Where they are convergent, state their sum to infinity.

(i) $100+99+98+97+\ldots$

(ii) $100+50+25+12\frac{1}{2}+\ldots$

(iii) $1-1+1-1+1-1+1-\ldots$

(iv) $4-2+1-\frac{1}{2}+\frac{1}{4}-\ldots$

(v) $0+0+0+0+0+0+\ldots$

(vi) $1/10+1/10+1/10+1/10+1/10+\ldots$

2. Explain why the following series are divergent:

(i) $\sum \dfrac{n}{2n+1}$. (ii) $\sum \sin\left(\dfrac{n\pi}{6}\right)$.

3. Why does Theorem 4.2.4 tell us nothing about the convergence or divergence of the series

$$1+\tfrac{1}{4}+\tfrac{1}{9}+\tfrac{1}{16}+\tfrac{1}{25}+\tfrac{1}{36}+\tfrac{1}{49}\ldots?$$

4. Investigate, for all values of a and d, the behaviour, as n tends to infinity, of the sequence $\{b_n\}$ defined by $b_n = dn^2/2+(a-d/2)n$. Justify your results. (See Theorem 2.4.1.)

4.3. More series, convergent and divergent

THEOREM 4.3.1. *The series $\Sigma 1/n^P$ is convergent if $P>1$ and divergent if $P \leqslant 1$.*

Proof. When $P=1$, $\Sigma 1/n^P$ becomes $\Sigma 1/n$, the harmonic series, which is divergent (Theorem 4.2.3).

Suppose $P<1$. Let the nth partial sums of the series $\Sigma 1/n^P$ and $\Sigma 1/n$ be S_n and T_n respectively. Since, for all n, $1/n^P \geqslant 1/n$, then $S_n \geqslant T_n$ for all n. Since the harmonic series is divergent, $\lim T_n = \infty$. Thus, by Theorem 3.3.7, $\lim S_n = \infty$. Therefore $\Sigma 1/n^P$ is divergent.

Now suppose $P>1$. Let the nth partial sum of the series be S_n.

$$S_1 = 1,$$

$$S_3 = 1+\frac{1}{2^P}+\frac{1}{3^P} < 1+\frac{1}{2^P}+\frac{1}{2^P} = 1+\frac{1}{2^{P-1}},$$

$$S_7 = 1+\left(\frac{1}{2^P}+\frac{1}{3^P}\right)+\left(\frac{1}{4^P}+\frac{1}{5^P}+\frac{1}{6^P}+\frac{1}{7^P}\right) < 1+\frac{1}{2^{P-1}}+\frac{1}{4^{P-1}}$$

$$= \frac{1-\left(\dfrac{1}{2^{P-1}}\right)^3}{1-\dfrac{1}{2^{P-1}}} < \frac{1}{1-\dfrac{1}{2^{P-1}}},$$

$$S_{15} < 1+\frac{1}{2^{P-1}}+\frac{1}{4^{P-1}}+\frac{1}{8^{P-1}} = \frac{\left(1-\dfrac{1}{2^{P-1}}\right)^4}{1-\dfrac{1}{2^{P-1}}} < \frac{1}{1-\dfrac{1}{2^{P-1}}}.$$

Continuing in this way, the (2^n-1)th partial sum is less than

$$\frac{1}{1-\dfrac{1}{2^{P-1}}}$$

for all n. But $\{S_n\}$ is an increasing sequence, since all the terms of the series are positive. Thus

$$S_n < \frac{1}{1-\dfrac{1}{2^{P-1}}}$$

for all n. Therefore $\{S_n\}$ is an increasing, bounded sequence, and is thus convergent by Theorem 3.4.1. Hence $\Sigma\, 1/n^P$ is convergent. □

The theorem tells us, for example, that the series

$$1+\tfrac{1}{4}+\tfrac{1}{9}+\tfrac{1}{16}+\tfrac{1}{25}+\tfrac{1}{36}+\tfrac{1}{49}+\tfrac{1}{64}+\dots$$

is convergent, whereas the series

$$1+\frac{1}{\sqrt{2}}+\frac{1}{\sqrt{3}}+\frac{1}{\sqrt{4}}+\dots$$

is divergent. We note that there is a great difference between being able to prove that a series is convergent and being able to find its sum to infinity. The sum to infinity of the series

$$1+\tfrac{1}{4}+\tfrac{1}{9}+\tfrac{1}{16}+\tfrac{1}{25}+\tfrac{1}{49}+\dots$$

is, in fact, $\pi^2/6$, but we are at present far from being able to obtain this result.

Series whose nth term is a difference. Consider the series

$$\frac{1}{1.2}+\frac{1}{2.3}+\frac{1}{3.4}+\frac{1}{4.5}+\dots.$$

$$a_n = \frac{1}{n(n+1)} = \frac{1}{n}-\frac{1}{n+1}.$$

Hence we may write the series

$$\left(1-\frac{1}{2}\right)+\left(\frac{1}{2}-\frac{1}{3}\right)+\left(\frac{1}{3}-\frac{1}{4}\right)+\left(\frac{1}{4}-\frac{1}{5}\right)+\ldots+\left(\frac{1}{n}-\frac{1}{n+1}\right)+\ldots$$

and it can be seen that $S_n = 1-1/(n+1)$ so that $\lim S_n = 1$. Thus

$$\sum_{n=1}^{\infty} \frac{1}{n(n+1)}$$

is convergent and its sum to infinity is 1.

Exercises 4.3

1. Which of the following series are convergent?

(i) $1+\frac{1}{2\sqrt{2}}+\frac{1}{3\sqrt{3}}+\frac{1}{4\sqrt{4}}+\frac{1}{5\sqrt{5}}+\ldots$

(ii) $1+\frac{\sqrt[3]{2}}{2}+\frac{\sqrt[3]{3}}{3}+\frac{\sqrt[3]{4}}{4}+\frac{\sqrt[3]{5}}{5}+\ldots$

2. Show that the following series are convergent and find their sum to infinity:

(i) $\frac{1}{1.3}+\frac{1}{2.4}+\frac{1}{3.5}+\frac{1}{4.6}+\frac{1}{5.7}+\frac{1}{6.8}+\ldots$

(ii) $\frac{1}{1.2.3}+\frac{1}{2.3.4}+\frac{1}{3.4.5}+\frac{1}{4.5.6}+\ldots$

(Write the nth term in the form $\frac{1}{2}\left[\frac{1}{n(n+1)}-\frac{1}{(n+1)(n+2)}\right]$.)

(iii) $\frac{2}{1.3.5}+\frac{4}{3.5.7}+\frac{6}{5.7.9}+\ldots$

4.4 The comparison test

THEOREM 4.4.1. *Σa_n is a series of* positive terms *(i.e. $a_n \geqslant 0$ for all n). Σa_n is convergent if and only if its partial sums are bounded.*

Proof. Let the nth partial sum of Σa_n be S_n. $S_{n+1}-S_n = a_{n+1} \geqslant 0$ for all n. Therefore the sequence $\{S_n\}$ is increasing.

By Theorem 3.4.1, if $\{S_n\}$ is bounded, then it is convergent; if not, then $\lim S_n = \infty$.

Therefore Σa_n is convergent if and only if its partial sums are bounded. □

So far we have looked at a number of series and we have decided whether they are convergent or not by applying the definition of convergence. In practice it is not always convenient to apply this definition; so we devise tests which become criteria by which the convergence of a series may be judged. Such a test is the comparison test, which enables us to decide upon the convergence of a series of *positive terms* by comparing it with another series of positive terms whose behaviour is already known.

THEOREM 4.4.2. (the comparison test). *Σa_n and Σb_n are both series of* positive terms. *If $a_n \geqslant b_n \geqslant 0$ for all n, and if Σa_n is convergent, then so is Σb_n.*

Proof. Let the nth partial sums of Σa_n and Σb_n be S_n and T_n respectively. Then $T_n \leqslant S_n$ for all n. But $\{S_n\}$ is convergent and therefore bounded. Thus $\{T_n\}$ is bounded, and therefore, by Theorem 4.4.1, Σb_n is convergent. □

We illustrate the use of the comparison test with two examples.

We have to discover whether $\Sigma 1/(2^n+1)$ converges. We note first that we have a series of positive terms. We know that $\Sigma 1/2^n$ converges, since this is a geometric progression with common ratio $\frac{1}{2}$ (Theorem 4.2.2).

Also
$$0 < \frac{1}{2^n+1} < \frac{1}{2^n}.$$

Therefore, applying the comparison test, $\Sigma 1/(2^n+1)$ converges.

We now examine the series $\Sigma 1/(n \sin^2 n)$. We again note that we have a series of positive terms. We know that the harmonic series $\Sigma 1/n$ diverges.

Now
$$0 < \frac{1}{n} \leqslant \frac{1}{n \sin^2 n} \quad (\text{since } |\sin n| \leqslant 1).$$

If $\Sigma 1/(n \sin^2 n)$ were convergent, the comparison test would make $\Sigma 1/n$ convergent, which is a contradiction. Hence $\Sigma 1/(n \sin^2 n)$ is divergent.

Suppose we have to consider the convergence of the series $\sum_{n=2}^{\infty} 1/(n^2-\sqrt{n})$

We already know that the series $\sum_{n=2}^{\infty} 1/n^2$ is convergent. We also know that, when n is large, $\sqrt{n}$ is small in comparison with n^2, so that $1/(n^2-\sqrt{n})$ does not differ greatly from $1/n^2$. We cannot, however, use the comparison test as it has already been stated to compare these series, since $1/(n^2-\sqrt{n})$ is greater rather thans maller than $1/n^2$. We now state and prove a useful modification of the comparison test, which enables us to compare series like these.

THEOREM 4.4.3. (the limit form of the comparison test). *Σa_n and Σb_n are two series of strictly positive terms (i.e. $a_n > 0$ and $b_n > 0$ for all n). If $\lim a_n/b_n = l$ and $l \neq 0$, then the series Σa_n and Σb_n either both converge or both diverge.*

Proof. Suppose first that Σb_n is convergent. Since $\{a_n/b_n\}$ is convergent, by Theorem 3.3.2 it is bounded, i.e. there is a K with $a_n/b_n \leqslant K$ for all n. Therefore $0 < a_n \leqslant Kb_n$.

Σb_n is convergent. Thus, by Theorem 4.2.5, ΣKb_n is convergent. Therefore, by Theorem 4.4.2, Σa_n is convergent.

Now suppose Σa_n is convergent. Since $\lim \{a_n/b_n\} = l$, and $l \neq 0$, $\{b_n/a_n\}$ is convergent by the corollary to Theorem 3.3.3. We continue as in the first part. □

We now return to the series $\sum_{n=2}^{\infty} 1/(n^2-\sqrt{n})$. Both this series and the series $\sum_{n=2}^{\infty} 1/n^2$ are series of strictly positive terms, and $\sum_{n=2}^{\infty} 1/n^2$ is convergent.

$$\lim \left[\frac{1}{n^2-\sqrt{n}} \div \frac{1}{n^2}\right] = \lim \left[\frac{n^2}{n^2-\sqrt{n}}\right] = \lim \left[\frac{1}{1-\dfrac{1}{n\sqrt{n}}}\right] = 1.$$

Therefore $\sum_{n=0}^{\infty} 1/(n^2-\sqrt{n})$ is convergent by Theorem 4.4.3. □

THEOREM 4.4.4. *Changing a finite number of terms of a series does not affect the convergence or divergence of the series. (It does affect the sum to infinity of a convergent series, as is evident.)*

Proof. Let S_n be the nth partial sum of the series before its terms are changed and let T_n be the nth partial sum of the series after its terms are changed. Then for all sufficiently large n, $T_n = S_n + c$, where c is some constant. Thus by Theorem 3.3.3 modified in view of Theorem 3.3.9, $\{T_n\}$ is convergent if and only if $\{S_n\}$ is convergent. Also $\lim T_n = \lim S_n + c$, so that, unless c happens to be zero, the sum to infinity is changed. □

Exercises 4.4

1. State whether each of the following series is convergent or divergent; prove your assertions:

(i) $\sum_{n=1}^{\infty} \frac{1}{n^3+9}$.

(ii) $\sum_{n=1}^{\infty} \left(\frac{\cos n}{n}\right)^2$.

(iii) $\sum_{n=1}^{\infty} \frac{1}{\sqrt{n}+50}$.

(iv) $\sum_{n=1}^{\infty} \frac{n^2}{n^3\sqrt{n}-n+5}$.

(v) $\sum_{n=1}^{\infty} \frac{n+\sqrt{n}}{2n\sqrt{n}-3}$.

(vi) $\sum_{n=1}^{\infty} \frac{1}{2^n+3^n}$.

(vii) $\sum_{n=1}^{\infty} \left(\cos^n \frac{\pi}{6} + \sin^n \frac{\pi}{6}\right)$.

4.5. Decimal representation

We are now in a position to discuss a topic which is first presented at a relatively early stage in a child's mathematical education but which we have had to hold over from Chapter 1 so that an adequate account may be given.

Terminating decimals. If integers are written in base 10, a natural way of writing some fractions is to use columns to the right of the units column to represent tenths, hundredths, thousandths, and so on.

Thus if a rational is expressible in a form where its denominator is a power of 10, then it may be written as a (terminating) decimal. For example, $1/2 = 5/10$ may be written 0.5, $17/25 = 68/100$ may be written 0.68, $5/8 = 625/1000$ may be written 0.625.

Conversely, any terminating decimal will represent a rational expressible in a form where its denominator is a power of 10. Thus a rational is expressible as a recurring decimal if and only if its denominator is of the form $2^p 5^q$.

Recurring decimals. If a rational is such that its denominator must necessarily have a prime factor other than 2 or 5, then it is not possible to write it as a terminating decimal. For example, 1/3, 4/7, and 2/15 may not be so expressed. A "decimal" presentation of such numbers is nevertheless often useful; so we look for an infinite series of the form

$$\frac{a_1}{10}+\frac{a_2}{10^2}+\frac{a_3}{10^3}+\ldots+\frac{a_n}{10^n}+\ldots,$$

where a_n is an integer satisfying $0 \leqslant a_n \leqslant 9$, whose sum to infinity is equal to the required rational.

In fact we discover, not only that we are always able to find such an infinite series, but also that the pattern of integers $\{a_n\}$ is always a recurring one. Before we can prove this we require a result from number theory.

THEOREM 4.5.1. *If p and q are relatively prime natural numbers (i.e. they have no common factor other than* 1), *then we may find natural numbers m and k such that*

$$p^m = kq+1.$$

Proof. Suppose that the theorem is not true, i.e. suppose that p^m does not leave a remainder of 1 when divided by q for any m. Given any two natural numbers r and s, with $r > s$,

$$p^r - p^s = p^s(p^{r-s} - 1).$$

By our assumption $p^{r-s} - 1$ is not divisible by q. Thus, since p and q have no common factor, $p^s(p^{r-s} - 1)$ is not divisible by q. Therefore p^r and p^s leave different remainders when divided by q. But this is true for *all* r and s, i.e. every power of p leaves a different remainder when divided by q. This is evidently absurd, since there are only $q - 1$ different remainders on division by q. Hence the theorem is true. □

THEOREM 4.5.2. *Every positive rational is expressible in the form*

$$\frac{r}{10^n} + \frac{s}{10^n(10^m - 1)},$$

where m is a natural number, n, r, and s are natural numbers or zero, and $s < 10^m - 1$.

Proof. If a rational has a denominator whose only prime factors are 2 and 5, then it is expressible in the form $r/10^n$.

Otherwise it is expressible as $p/10^n q$, where q and 10 are relatively prime. According to Theorem 4.5.1, we may find natural numbers m and k such that $kq = 10^m - 1$.

Thus
$$\frac{p}{10^n q} = \frac{kp}{10^n kq} = \frac{kp}{10^n(10^m - 1)}.$$

Now we may find r and s, with $s < 10^m - 1$, such that

$$kp = r \times 10^{m-1} + s.$$

Then
$$\frac{p}{10^n q} = \frac{kp}{10^n(10^m - 1)} = \frac{r}{10^n} + \frac{s}{10^n(10^m - 1)}. \quad \square$$

For example,

$$\frac{1}{6} = \frac{15}{10 \times 9}; \quad \frac{53}{275} = \frac{53}{25 \times 11} = \frac{1908}{100 \times 99} = \frac{19}{100} + \frac{27}{100 \times 99}.$$

THEOREM 4.5.3. *Every positive rational is expressible in one of the forms*

$$\text{(i) } \frac{r}{10^n} \quad \text{or} \quad \text{(ii) } \frac{r}{10^n}+\frac{1}{10^{n+m}}\left(s+\frac{s}{10^m}+\frac{s}{10^{2m}}+\frac{s}{10^{3m}}+\ldots\right)$$

where m, n, r, and s satisfy the conditions of Theorem 4.5.2, *i.e. every positive rational is expressible as a terminating or recurring decimal respectively. Conversely, every positive terminating or recurring decimal represents a positive rational.*

Proof. Suppose we are given a rational. If, when it is expressed in the form of Theorem 4.5.2, s is zero, then the rational is of the form (i) and may be represented by a terminating decimal. Otherwise, by Theorem 4.2.2,

$$\frac{s}{10^n(10^m-1)} = \frac{1}{10^{n+m}}\left(s+\frac{s}{10^m}+\frac{s}{10^{2m}}+\frac{s}{10^{3m}}+\ldots\right).$$

Thus the rational is expressible in the form (ii) and may be represented as a recurring decimal.

Conversely, any positive recurring decimal is expressible in the form (ii) and every terminating decimal in the form (i), and hence they are expressible in the form of Theorem 4.5.2 and represent rationals. □

We have for convenience restricted our remarks to positive rationals. Evidently, negative rationals introduce no further complications and may also be expressed as terminating or recurring decimals.

Irrationals and decimals. We now seek to represent irrationals as decimals. In view of Theorem 4.5.3, such decimals will *not* be terminating or recurring decimals.

THEOREM 4.5.4. *Every irrational may be expressed as a non-recurring infinite decimal, i.e. given an irrational x, we may find integers n, a_1, a_2,*

$a_3, a_4, \ldots$, with $n \geqslant 0$ and $0 \leqslant a_r \leqslant 9$ for all r, such that x is the sum to infinity of the series

$$\pm\left(n+\frac{a_1}{10}+\frac{a_2}{10^2}+\frac{a_3}{10^3}+\frac{a_4}{10^4}+\ldots\right).$$

Conversely every infinite non-recurring decimal represents an irrational.

Proof. Suppose, without loss of generality, that x is positive. Let n be the greatest integer satisfying $n < x$. Then $x-n < 1$. Choose a_1 so that $n+a_1 10^{-1} < x$ but $n+(a_1+1)\, 10^{-1} > x$. Then $x-(n+a_1 10^{-1}) < 10^{-1}$. Now choose a_2 so that $n+a_1\, 10^{-1}+a_2\, 10^{-2} < x$ but $n+a_1 10^{-1}+(a_2+1)\, 10^{-2} > x$. Then $x-(n+a_1\, 10^{-1}+a_2\, 10^{-2}) < 10^{-2}$. Continuing in this way, we define $a_3, a_4, a_5, \ldots$, and obtain a series

$$n+a_1\, 10^{-1}+a_2\, 10^{-2}+a_3\, 10^{-3}+\ldots$$

of positive terms, whose $(n+1)$th partial sum S_{n-1} satisfies $0 < x-S_{n-1} < 10^{-n}$. We deduce, by Theorem 3.3.8, that $\lim(x-S_{n+1}) = 0$. Thus x is the sum to infinity of the infinite series

$$n+\frac{a_1}{10}+\frac{a_2}{10^2}+\frac{a_3}{10^3}+\ldots.$$

Conversely, if we are given such a series, then every term is less than or equal to the corresponding term of the series

$$n+\frac{9}{10}+\frac{9}{10^2}+\frac{9}{10^3}+\ldots$$

which is convergent, since it is a geometric progression with common ratio 1/10. Hence, by the comparison test, the series

$$n+\frac{a_1}{10}+\frac{a_2}{10^2}+\frac{a_3}{10^3}+\ldots$$

is convergent. Its sum to infinity x must be irrational in view of Theorem 4.5.3. □

If we are given an irrational, such as $\sqrt{2}$ or π, then we may obtain, by means of a suitable process, the first few terms of its expression as an infinite decimal. Thus

$$\sqrt{2} = 1.414\ldots,$$
$$\pi = 3.1415926535897\ldots.$$

In these cases there is no simple rule satisfied by the nth digit of the expansion.

On the other hand, we may construct infinite decimals representing irrationals by defining the nth digit of the expansion according to a rule provided that the rule does not give a recurring pattern. For example, the expansion

$$0.1101000100000001\ldots$$

represents an irrational if the nth digit is defined to be 1 when n is a power of 2 and zero otherwise. Such a method of construction does not generally give irrationals which are familiar in other contexts.

Summary. We have seen that every rational number is expressible as a terminating or as a recurring decimal and conversely; also that every irrational number is expressible as a non-recurring infinite decimal and conversely.

Exercises 4.5

1. Write the recurring decimal $1.\dot{3}12\dot{4}$ as an infinite series and hence express it as a rational.

2. Obtain the decimal presentations of 1/7, 2/7, 3/7, 4/7, 5/7, and 6/7. Comment. Explain why this phenomenon arises. Why does it not arise with 1/11, 2/11, 3/11, 4/11 ...?

3. Obtain by trial and error the first three digits in the decimal presentation of $\sqrt[3]{(1/2)}$.

4. Give three examples of decimals for which the nth digit is defined by some simple rule but which represent irrationals.

5. Express 2/7, 1/11, 3/13, 2/17, 14/19 as recurring decimals. Postulate some connection between the number of digits in the recurring pattern and the denominator of the rational, when the denominator is a prime other than 2 or 5. Prove your assertion.

4.6. Absolute convergence

So far we have mainly been concerned with series of positive terms. We now wish to focus attention more specifically on series whose terms are mixed in sign, some being positive and some negative. Consider, for example, the series

$$1-\tfrac{1}{4}+\tfrac{1}{9}-\tfrac{1}{16}+\tfrac{1}{25}-\tfrac{1}{36}+\tfrac{1}{49}-\tfrac{1}{64}+\dots.$$

In order to investigate the convergence of such a series, it is often convenient to consider the convergence of a related series; in this case the series

$$1+\tfrac{1}{4}+\tfrac{1}{9}+\tfrac{1}{16}+\tfrac{1}{25}+\tfrac{1}{36}+\tfrac{1}{49}+\tfrac{1}{64}+\dots$$

the second series being obtained from the first by replacing all the negative signs by positive signs (or, in other words, by replacing every term by its modulus). Since the second series is convergent, we shall describe the first series as absolutely convergent.

DEFINITION. *Given a series Σa_n, if $\Sigma |a_n|$ is convergent then we say that Σa_n is* absolutely convergent.

To see whether a series is absolutely convergent we test, for convergence, not the series itself but another series. Therefore if a series is absolutely convergent we do not yet know that it is convergent. All absolutely convergent series *are* convergent, however, as the following theorem shows.

THEOREM 4.6.1. *If a series is absolutely convergent, then it is convergent.*

Proof. Suppose that our series is Σa_n. Then we know that $\Sigma |a_n|$ is convergent. Consider the series Σb_n defined by

$$b_n = a_n \quad \text{if } a_n \geqslant 0,$$
$$b_n = 0 \quad \text{if } a_n < 0.$$

We have $0 \leqslant b_n \leqslant |a_n|$ for all n. Since $\Sigma |a_n|$ is convergent, we deduce from the comparison test that Σb_n is convergent. Consider now the series Σc_n defined by

$$c_n = 0 \quad \text{if } a_n \geqslant 0,$$
$$c_n = -a_n \quad \text{if } a_n < 0.$$

We have $0 \leqslant c_n \leqslant |a_n|$ for all n. Since $\Sigma |a_n|$ is convergent, we deduce from the comparison test that Σc_n is convergent. Therefore, by Theorem 4.2.5, $\Sigma (-c_n)$ is convergent. But $a_n = b_n - c_n$ for all n. Thus, by Theorem 4.2.5, Σa_n is convergent. □

For example, the first series of this section is absolutely convergent and is now, therefore, convergent. Or consider the series

$$\tfrac{1}{3} - \tfrac{1}{5} + \tfrac{1}{9} - \tfrac{1}{17} + \tfrac{1}{33}$$

whose nth term is $(-1)^{n+1}/(2^n+1)$. We prefer to discuss the absolute convergence of this series since we may then use the comparison test.

Since $0 < 1/(2^n+1) < 1/2^n$ for all n and since $\Sigma\, 1/2^n$ is convergent, then by the comparison test $\Sigma\, 1/(2^n+1)$ is convergent. Thus the series $\Sigma (-1)^{n+1}/(2^n+1)$ is absolutely convergent, and so, by Theorem 4.6.1, it is convergent.

Tests for absolute convergence. We now give two tests for the absolute convergence of a series. Both of these tests are based upon the behaviour of geometric progressions; both are relatively easy to apply.

THEOREM 4.6.2. (D'Alembert's ratio test). *Σa_n is a series in which $a_n \neq 0$ for all n.*

(i) *If* $\lim |a_{n+1}/a_n| = l$ *and* $l < 1$, *then* Σa_n *is absolutely convergent.*

(ii) *If* $\lim |a_{n+1}/a_n| = l$ *and* $l > 1$, *then* Σa_n *is divergent.*

(iii) *If* $\lim |a_{n+1}/a_n| = 1$, *then* Σa_n *may be either convergent or divergent (in other words the test fails).*

Proof. (i) Put $r = (l+1)/2$ so that $l < r < 1$. Since $\lim |a_{n+1}/a_n| = l$, there is a natural number N such that, for all $n \geqslant N$, $||a_{n+1}/a_n| - l| < r - l$.

Hence $|a_{n+1}/a_n| < r$.

According to Theorem 4.4.4 we may discard the first $N-1$ terms of the series without affecting its convergence or divergence. So we consider $\sum\limits_{n=N}^{\infty} a_n$.

Since $|a_{N+1}/a_N| < r$ and $|a_{N+2}/a_{N+1}| < r$ we have $|a_{N+2}/a_N| < r^2$. Continuing in this way we have $|a_{N+n}/a_N| < r^n$. Thus $0 < |a_{N+n}| < |a_N|\, r^n$. Now $\Sigma |a_N|\, r^n$ is convergent since it is a geometric progression with common ratio $r < 1$ (Theorem 4.2.2). Thus, by the comparison test, $\sum\limits_{n=N}^{\infty} |a_n|$ is convergent and so $\sum\limits_{n=0}^{\infty} a_n$ is convergent.

(ii) If $l > 1$ then $l-1 > 0$. So we can find a natural number N such that, for all $n \geqslant N$, $||a_{n+1}/a_n| - l| < l-1$, and so $|a_{n+1}/a_n| > 1$. Therefore $|a_{N+n}| \geqslant |a_N| > 0$ for all $n \geqslant 0$. Thus, by Theorem 3.3.6, $\lim a_n \neq 0$. Therefore, by Theorem 4.2.4, Σa_n is divergent.

(iii) (a) If $a_n = 1/n$, then $\lim |a_{n+1}/a_n| = \lim [n/(n+1)] = 1$.
Σa_n is divergent (Theorem 4.2.3).

(b) If $a_n = 1/n^2$, then $\lim |a_{n+1}/a_n| = \lim [n^2/(n+1)^2] = 1$.
Σa_n is convergent (Theorem 4.3.1). □

To illustrate the use of D'Alembert's test, we investigate the series $\sum\limits_{n=1}^{\infty} x^n/n^2$ for different real values of x. Put $a_n = x^n/n^2$. If $x = 0$, then the series is trivially convergent. If $x \neq 0$, then $\lim |a_{n=1}/a_n| = \lim [|x|n^2/(n+1)^2] = |x|$. Therefore if $|x| < 1$ the series is absolutely convergent by D'Alembert's ratio test. If $|x| > 1$, the series is divergent by D'Alembert's ratio test. If $|x| = 1$, D'Alembert's ratio test fails. However, by

Theorem 4.3.1, $\Sigma 1/n^2$ is convergent, so that our series is absolutely convergent when $|x| = 1$.

D'Alembert's ratio test fails to establish the convergence and divergence of every series, as we have seen. Cauchy's test also fails sometimes, but not so often. In fact Cauchy's test works whenever D'Alembert's test works, and in other cases as well.

THEOREM 4.6.3. (Cauchy's test). (i) *If* $\overline{\lim}\, |a_n|^{1/n} = \Lambda < 1$, *then* $\sum_{n=1}^{\infty} a_n$ *is absolutely convergent.*

(ii) *If* $\overline{\lim}\, |a_n|^{1/n} = \Lambda > 1$, *then* $\sum_{n=1}^{\infty} a_n$ *is divergent.*

Proof. (i) Put $r = (1+\Lambda)/2$ so that $\Lambda < r < 1$. By Theorem 3.7.3 we can find a natural number N such that, for all $n \geqslant N$, $|a_n|^{1/n} < \Lambda + (r-\Lambda) = r$. As in the proof of D'Alembert's test, we discard the first $N-1$ terms and consider $\sum_{n=N}^{\infty} a_n$. For all $n \geqslant N, 0 < |a_n| < r^n$. $\sum_{n=N}^{\infty} r_n$ is convergent, since it is a geometric progression with common ratio $r < 1$. Therefore, by the comparison test, $\sum_{n=N}^{\infty} a_n$ is absolutely convergent and so $\sum_{n=1}^{\infty} a_n$ is absolutely convergent.

(ii) By Theorem 3.7.3 $a_n > \Lambda - (\Lambda - 1)$ for some n as large as we please. Therefore $a_n > 1$ for some n as large as we please. Therefore $\lim a_n \neq 0$ (Theorem 3.3.6). Thus, by Theorem 4.2.6, Σa_n is divergent. □

Consider the following series as an example where Cauchy's test succeeds but D'Alembert's test fails:

$$\tfrac{1}{3} + \tfrac{2}{3} + \tfrac{2}{9} + \tfrac{4}{9} + \tfrac{4}{27} + \tfrac{8}{27} + \dots$$

If the nth term of this series is a_n, then

$$a_{2n} = (\tfrac{1}{3})^n 2^n,$$
$$a_{2n+1} = (\tfrac{1}{3})^{n+1} 2^n.$$

Thus $\lim [(a_n)^{1/n}] = \sqrt{\tfrac{2}{3}}$. So, by Cauchy's test, the series is convergent.

However, since $a_{2n}/a_{2n-1} = 2$ for all n, $a_{2n+1}/a_{2n} = \frac{1}{3}$ for all n, D'Alembert's test fails.

In Chapter 2 we proved the general principle of convergence for sequences, which enables us to test whether a sequence is convergent without needing to know its limit.

We now give this result as it applies to series.

THEOREM 4.6.4. (the general principle of convergence for series). *Σa_n is convergent if and only if given $\varepsilon > 0$ we can find a natural number N such that $\left|\sum_{n=P}^{Q} a_n\right| < \varepsilon$ for all $Q > P \geqslant N$.*

Proof. Let S_n be the nth partial sum of Σa_n. Then $\sum_{n=P}^{Q} a_n = S_Q - S_{P-1}$ and the theorem follows immediately from the definition of a Cauchy sequence and Theorem 3.7.5.

THEOREM 4.6.5. (the general principle of absolute convergence for series). *Σa_n is absolutely convergent if and only if given $\varepsilon > 0$ we can find a natural number N such that $\sum_{n=P}^{Q} |a_n| < \varepsilon$ for all $Q > P \geqslant N$.*

Proof. This theorem follows immediately from the definition of absolute convergence and Theorem 4.6.4.

Exercises 4.6

1. Show that $\sum \left(-\frac{1}{2}\right)^n \frac{4.\,8.\,12.\,16.\,\ldots\,.\,4n}{3.\,7.\,11.\,15.\,\ldots\,.\,4n-1}$ is absolutely convergent.

2. For what values of x does each of the following series converge:

(i) $\sum \frac{x^n}{n}$.

(ii) $\sum \frac{x^n}{n^3}$.

(iii) $\sum nx^n$

(iv) $\sum \frac{n^2}{x^n}$.

(v) $\sum n!\, x^n$.

(vi) $\sum n!\, x^{n^2}$.

3. Give an example of a series whose terms are monotonic decreasing which may be proved convergent by Cauchy's test but not by D'Alembert's test.

4. Criticize the use of D'Alembert's test to prove that Σx^n is convergent when $|x| < 1$.

5. Show that, if a series may be proved convergent by D'Alembert's test, then it may be proved convergent by Cauchy's test.

4.7. Conditional convergence

In the last section we saw that one way of dealing with a series containing both positive and negative terms is to test it for absolute convergence or, in other words, to examine for convergence the corresponding series all of whose terms are positive. In this section we discuss the convergence of series which contain positive and negative terms and which are not absolutely convergent.

Consider, for example, the series

$$1-\tfrac{1}{2}+\tfrac{1}{3}-\tfrac{1}{4}+\tfrac{1}{5}-\tfrac{1}{6}+\tfrac{1}{7}-\ldots.$$

This series is not absolutely convergent. That is, the series

$$1+\tfrac{1}{2}+\tfrac{1}{3}+\tfrac{1}{4}+\tfrac{1}{5}+\tfrac{1}{6}+\tfrac{1}{7}+\ldots$$

(the harmonic series) is divergent. However, we shall see later that the first series is, nevertheless, convergent.

DEFINITION. *A series which is convergent but not absolutely convergent is called* conditionally *convergent.*

Thus, before we can assert that a series is conditionally convergent, we must prove two facts about it; first that it is convergent, and, second, that it is not absolutely convergent.

We observe that the terms of the first series of this section alternate strictly, i.e. every positive term is immediately followed by a negative term and vice versa. Such a series is called an alternating series.

THEOREM 4.7.1. (the alternating series test). *The series $\Sigma(-1)^{n+1}a_n$ is convergent if the following conditions are all satisfied:*

(i) $a_n > 0$ *for all n (i.e. the series alternates),*

(ii) $\{a_n\}$ *is a monotonic decreasing sequence,*

(iii) $\lim a_n = 0$.

Proof. Let S_n be the nth partial sum of the series.

Then $$S_{2n} = S_{2n-2}+a_{2n-1}-a_{2n} > S_{2n-2}.$$

Also $$S_{2n} = a_1-(a_2-a_3)-(a_4-a_5)-(a_6-a_7)-\dots$$
$$-(a_{2n-2}-a_{2n-1})-a_{2n} < a_1.$$

Therefore $\{S_{2n}\}$ is an increasing sequence which is bounded above by a_1. Therefore, by Theorem 3.4.1, $\{S_{2n}\}$ is convergent. Let $\lim S_{2n} = l$. $S_{2n+1} = S_{2n}+a_{2n+1}$: But $\lim S_{2n} = l$ and $\lim a_{2n+1} = 0$. Therefore, by Theorem 3.3.3, $\lim S_{2n+1} = \lim S_{2n} = l$. Therefore $\lim S_n = l$ and $\Sigma(-1)^{n+1}a_n$ is convergent. □

Consider again the first series of this section whose nth term is $(-1)^{n+1}\,1/n$. (i) $1/n > 0$ for all n. (ii) $\{1/n\}$ is monotonic decreasing. (iii) $\lim\,[1/n] = 0$. Therefore, by the alternating series test, the series $\Sigma(-1)^{n+1}\,1/n$ is convergent. Moreover, since the harmonic series is divergent, the series $\Sigma(-1)^{n+1}\,1/n$ is conditionally convergent.

Other conditionally convergent series. The alternating series test is just one test for the convergence of conditionally convergent series, although it is the only such test which we shall consider. Series which do not satisfy the conditions of the alternating series test may still be conditionally convergent. The reader will be able to show that

$$1-\tfrac{1}{2}-\tfrac{1}{2}+\tfrac{1}{2}-\tfrac{1}{4}-\tfrac{1}{4}+\tfrac{1}{3}-\tfrac{1}{6}-\tfrac{1}{6}+\tfrac{1}{4}-\tfrac{1}{8}-\tfrac{1}{8}+\dots$$

is such a series.

Exercises 4.7

1. Show that the series

$$1-\tfrac{1}{2}-\tfrac{1}{2}+\tfrac{1}{2}-\tfrac{1}{4}-\tfrac{1}{4}+\tfrac{1}{3}-\tfrac{1}{6}-\tfrac{1}{6}+\tfrac{1}{4}-\tfrac{1}{8}-\tfrac{1}{8}+\dots$$

is conditionally convergent. Explain why it does not satisfy the conditions of the alternating series test.

2. Show that the series

$$1-\tfrac{1}{4}+\tfrac{1}{9}-\tfrac{1}{16}+\tfrac{1}{25}-\tfrac{1}{36}+\tfrac{1}{49}-\tfrac{1}{64}+\dots$$

satisfies the conditions of the alternating series test and is therefore convergent. Explain why it is not conditionally convergent.

3. Say of each of the following series whether it is absolutely convergent, conditionally convergent or divergent:

(i) $\sum(-1)^{n+1}\dfrac{1}{n^5}$.

(ii) $\sum(-1)^{n+1}\cos\left(\dfrac{1}{n}\right)$.

(iii) $\sum(-1)^{n-1}\dfrac{1}{\sqrt{n}}$.

(iv) $\sum(-1)^{n+1}\dfrac{n+1}{2n\sqrt{n}}$.

4. Explain why each of the following series fails to satisfy the conditions of the alternating series test:

(i) $1-\tfrac{1}{3}+\tfrac{1}{2}-\tfrac{1}{4}+\tfrac{1}{3}-\tfrac{1}{5}+\tfrac{1}{4}-\tfrac{1}{6}+\tfrac{1}{5}-\tfrac{1}{7}+\dots$

(ii) $2-1\tfrac{1}{2}+1\tfrac{1}{4}-1\tfrac{1}{8}+1\tfrac{1}{16}-1\tfrac{1}{32}+\dots$

(iii) $1-\tfrac{1}{2}-\tfrac{1}{3}+\tfrac{1}{4}-\tfrac{1}{5}-\tfrac{1}{6}+\tfrac{1}{7}-\tfrac{1}{8}-\tfrac{1}{9}+\dots$

Show that the series (i) is convergent and that the series (ii) is divergent.

5. Prove that, if a series is conditionally convergent, then it must have an infinite number of both positive and negative terms.

Prove also that the subsequence of positive terms and the subsequence of negative terms are both divergent.

4.8. Rearrangement of series

The remainder of the chapter is devoted to results which contrast the behaviour of absolutely convergent series and conditionally convergent series. In this section we are concerned with the rearrangement of the order of terms in a series.

If we have a sum of a finite set of real numbers, then we know that the sum remains unchanged if we change the order of the terms in the summation. For example,

$$2-3+4-5+6 = 2+4+6-3-5 = 6-5+4-3+2, \text{ and so on.}$$

This is a consequence of the commutativity of addition for real numbers. At first sight it might be expected that this would apply also to infinite series. However, this is not always the case. To be exact, the order of the terms forming an infinite series may be changed without affecting the sum to infinity if and only if the infinite series is *absolutely* convergent. We prove this in the form of three theorems.

THEOREM 4.8.1. *If* $\sum_{n=1}^{\infty} a_n$ *is a convergent series of positive terms, its terms may be rearranged without altering its sum to infinity; i.e. if* $\sum_{n=1}^{\infty} b_n$ *is another series such that:*

(i) *for all* n *we can find an* m *with* $b_n = a_m$,

(ii) *for all* q *we can find an* r *with* $a_q = b_r$,

then $\sum_{n=1}^{\infty} b_n = \sum_{n=1}^{\infty} a_n$.

Proof. Let $\sum_{n=1}^{\infty} a_n = l$. Since $a_n \geqslant 0$ for all n, $\sum_{n=1}^{P} a_n \leqslant l$ and $\sum_{n=1}^{Q} b_n \leqslant l$ for all P and Q. But $\sum_{n=1}^{\infty} b_n$ is a series of positive terms. Therefore $\sum_{n=1}^{\infty} b_n$ satisfies the conditions of Theorem 4.4.1 and is convergent. Also, by Theorem 3.3.6, $\sum_{n=1}^{\infty} b_n \leqslant l$, i.e. $\sum_{n=1}^{\infty} b_n \leqslant \sum_{n=1}^{\infty} a_n$. So too, $\sum_{n=1}^{\infty} a_n \leqslant \sum_{n=1}^{\infty} b_n$. Thus $\sum_{n=1}^{\infty} a_n = \sum b_n$. □

THEOREM 4.8.2. *If* $\sum_{n=1}^{\infty} a_n$ *is an absolutely convergent series, its terms may be rearranged without altering its sum to infinity, i.e. if* $\sum_{n=1}^{\infty} b_n$ *is another series, satisfying the same conditions as in Theorem* 4.8.1, *then* $\sum_{n=1}^{\infty} b_n = \sum_{n=1}^{\infty} a_n$.

Proof. Define the series $\sum_{n=1}^{\infty} c_n$ and $\sum_{n=1}^{\infty} d_n$ by the equations

$$\begin{aligned} c_n &= a_n \quad \text{if} \quad a_n \geqslant 0, \quad & d_n &= 0 \quad \text{if} \quad a_n \geqslant 0, \\ c_n &= 0 \quad \text{if} \quad a_n < 0, \quad & d_n &= -a_n \quad \text{if} \quad a_n < 0. \end{aligned}$$

For all n, $0 \leqslant c_n \leqslant |a_n|$ and $0 \leqslant d_n \leqslant |a_n|$. But $\sum_{n=1}^{\infty} |a_n|$ is convergent. Therefore by the comparison test $\sum_{n=1}^{\infty} c_n$ and $\sum_{n=1}^{\infty} d_n$ are convergent. Since for all n $a_n = c_n - d_n$. Theorem 4.2.5 gives

$$\sum_{n=1}^{\infty} a_n = \sum_{n=1}^{\infty} c_n - \sum_{n=1}^{\infty} d_n.$$

Define the series $\sum_{n=1}^{\infty} C_n$ and $\sum_{n=1}^{\infty} D_n$ by the equations:

$$\begin{aligned} C_n &= b_n \quad \text{if} \quad b_n \geqslant 0, \quad & D_n &= 0 \quad \text{if} \quad b_n \geqslant 0, \\ C_n &= 0 \quad \text{if} \quad b_n < 0, \quad & D_n &= -b_n \quad \text{if} \quad b_n < 0. \end{aligned}$$

Then $\sum_{n=1}^{\infty} C_n$ and $\sum_{n=1}^{\infty} D_n$ are similarly convergent and

$$\sum_{n=1}^{\infty} b_n = \sum_{n=1}^{\infty} C_n - \sum_{n=1}^{\infty} D_n.$$

Now $\sum_{n=1}^{\infty} c_n$ is a series of positive terms and $\sum_{n=1}^{\infty} C_n$ is a rearrangement of $\sum_{n=1}^{\infty} c_n$. Therefore, by Theorem 4.8.1, $\sum_{n=1}^{\infty} c_n = \sum_{n=1}^{\infty} C_n$. So, too, $\sum_{n=1}^{\infty} d_n = \sum_{n=1}^{\infty} D_n$. Thus $\sum_{n=1}^{\infty} a_n = \sum_{n=1}^{\infty} b_n$. □

THEOREM 4.8.3. *If* $\sum_{n=1}^{\infty} a_n$ *is a conditionally convergent series, its terms may be rearranged in such a way that the rearranged series is convergent,* with any sum to infinity we please, *or is divergent.*

Proof. Some of the terms of the series are positive; others are negative. We shall call the nth positive term b_n and the nth negative term c_n.

Since $\sum_{n=1}^{\infty} a_n$ is conditionally convergent, $\sum_{n=1}^{\infty} b_n$ and $\sum_{n=1}^{\infty} c_n$ are both infinite series and are both divergent (see exercises 4.7, question 5).

We shall prove that the series $\sum_{n-1}^{\infty} a_n$ may be rearranged so that its sum to infinity is the real number $x > 0$. We leave the proofs of the other cases to the reader.

To begin our rearrangement we use b_1. We then use as many of the subsequent terms of $\sum_{n=1}^{\infty} b_n$ as are necessary for the partial sum first to exceed x. We then use c_1 and as many of the subsequent terms of $\sum_{n=1}^{\infty} c_n$ as are necessary for the partial sum to fall below x: we then continue adding terms from $\sum_{n=1}^{\infty} b_n$ until the partial sum again exceeds x. We then continue adding terms from $\sum_{n=1}^{\infty} c_n$ and so on. Since $\lim b_n = 0$ and $\lim c_n = 0$, the amount by which the partial sum exceeds x after adding terms from $\sum_{n=1}^{\infty} b_n$ and the amount by which the partial sum falls short of x after adding terms from $\sum_{n=1}^{\infty} c_n$ both tend to zero. Thus if S_n is the nth partial sum, $\overline{\lim}\, S_n = x$ and $\underline{\lim}\, S_n = x$. Therefore $\lim S_n = x$ and the rearranged series converges with x as its sum to infinity. □

Exercises 4.8

1. Find the sum to infinity of the series

$$1-\tfrac{1}{2}-\tfrac{1}{8}+\tfrac{1}{4}-\tfrac{1}{32}-\tfrac{1}{128}+\tfrac{1}{16}-\tfrac{1}{512}-\tfrac{1}{2048}+\ldots$$

2. Prove that a conditionally convergent series may be rearranged so that its sum to infinity is equal to the negative real number x.

3. Prove that a conditionally convergent series may be rearranged so that it is divergent.

4. Discuss the effect of changing the order of a finite number of the terms of a series.

5. If x is the sum to infinity of the series

$$1-\tfrac{1}{2}+\tfrac{1}{3}-\tfrac{1}{4}+\tfrac{1}{5}-\tfrac{1}{6}+\tfrac{1}{7}-\ldots$$

show that the rearrangement

$$1-\tfrac{1}{2}-\tfrac{1}{4}+\tfrac{1}{3}-\tfrac{1}{6}-\tfrac{1}{8}+\tfrac{1}{5}-\tfrac{1}{10}-\tfrac{1}{12}+\ldots$$

has sum to infinity $\tfrac{1}{2}x$.

4.9. Multiplication of series

If we have two finite sums, then we may multiply them together without difficulty and obtain the product of the sums. For example,

$$(a+bx+cx^2)(d+ex+fx^2)$$
$$= ad+(ae+bd)x+(af+be+cd)x^2+(bf+ce)x^3+cfx^4.$$

We now consider the multiplication of infinite series.

Suppose we have two convergent series $\sum_{n=0}^{\infty} a_n$ and $\sum_{n=0}^{\infty} b_n$. The set of all possible products of terms of the first series with terms of the second series may be conveniently set out in the form of an infinite matrix as in Fig. 4.1. If we multiply together the first terms of each series we obtain a_0b_0. If we multiply together the first two terms of each series we obtain $a_0b_0+(a_0b_1+a_1b_0+a_1b_1)$ and so on. At each stage we obtain

a_0b_0	a_0b_1	a_0b_2	a_0b_3	a_0b_4	a_0b_5	...
a_1b_0	a_1b_1	a_1b_2	a_1b_3	a_1b_4	a_1b_5	...
a_2b_0	a_2b_1	a_2b_2	a_2b_3	a_2b_4	a_2b_5	...
a_3b_0	a_3b_1	a_3b_2	a_3b_3	a_3b_4	a_3b_5	...
a_4b_0	a_4b_1	a_4b_2	a_4b_3	a_4b_4	a_4b_5	...
a_5b_0	a_5b_1	a_5b_2	a_5b_3	a_5b_4	a_5b_5	...
...	...	...	...	...	...	...
...	...	...	...	...	...	...

FIG. 4.1.

the sum of all the terms of a square submatrix, as indicated by the boxing in Fig. 4.1.

DEFINITION. *The* box product *of the series* $\sum_{n=0}^{\infty} a_n$ *and* $\sum_{n=0}^{\infty} b_n$ *is the series* $\sum_{n=0}^{\infty} d_n$, *where* d_n *is defined by*

$$d_n = (a_0+a_1+a_2+\ldots+a_{n-1})b_n+a_n(b_0+b_1+b_2+\ldots+b_{n-1})+a_nb_n.$$

THEOREM 4.9.1. *If* $\sum_{n=0}^{\infty} d_n$ *is the box product of the convergent series* $\sum_{n=0}^{\infty} a_n$ *and* $\sum_{u=0}^{\infty} b_n$, *then* $\sum_{n=0}^{\infty} d_n$ *is convergent and*

$$\sum_{n=0}^{\infty} d_n = \left(\sum_{n=0}^{\infty} a_n\right)\left(\sum_{n=0}^{\infty} b_n\right).$$

Proof. Let S_n, T_n, and U_n be the nth partial sums of $\sum_{n=0}^{\infty} a_n$, $\sum_{n=0}^{\infty} b_n$, and $\sum_{n=0}^{\infty} d_n$ respectively. Then $S_nT_n = U_n$ as may be readily seen from Fig. 4.1. Therefore, by Theorem 3.3.3, $\lim S_n \cdot \lim T_n = \lim U_n$, i.e.

$$\left(\sum_{n=0}^{\infty} a_n\right)\left(\sum_{n=0}^{\infty} b_n\right) = \sum_{n=0}^{\infty} d_n. \quad \square$$

In applications of the product of two series it is usually more convenient to use not the box product but another product called the Cauchy product. This is the product suggested by Fig. 4.2.

DEFINITION. *The* Cauchy product *of the series* $\sum_{n=0}^{\infty} a_n$ *and* $\sum_{n=0}^{\infty} b_n$ *is the series* $\sum_{n=0}^{\infty} c_n$, *where* c_n *is defined by*

$$c_n = a_0b_n+a_1b_{n-1}+a_2b_{n-2}+\ldots+a_{n-1}b_1+a_nb_0.$$

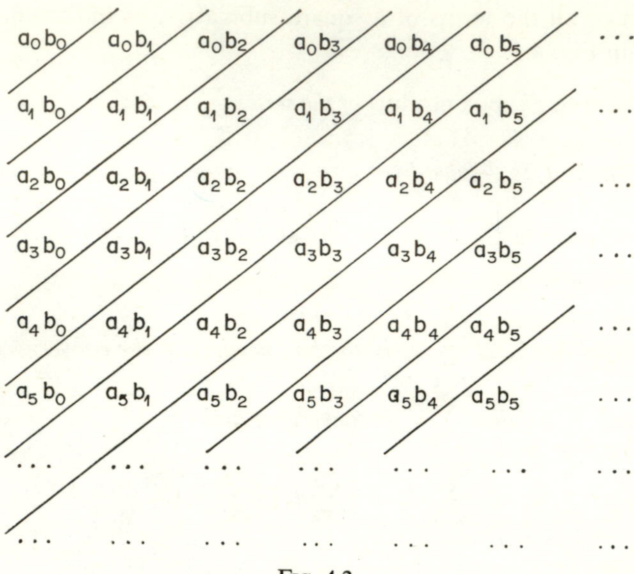

FIG. 4.2.

Since both the box product and the Cauchy product contain all possible products of the terms of one series with the terms of the other, the Cauchy product is a rearrangement of the box product.

THEOREM 4.9.2. *If* $\sum_{n=0}^{\infty} c_n$ *is the Cauchy product of the* absolutely *convergent series* $\sum_{n=0}^{\infty} a_n$ *and* $\sum_{n=0}^{\infty} b_n$, *then* $\sum_{n=0}^{\infty} c_n$ *is convergent and*

$$\sum_{n=0}^{\infty} c_n = \left(\sum_{n=0}^{\infty} a_n\right)\left(\sum_{n=0}^{\infty} b_n\right).$$

Proof. Let $\sum_{n=0}^{\infty} d_n$ be the box product of the series $\sum_{n=0}^{\infty} a_n$ and $\sum_{n=0}^{\infty} b_n$ and let $\sum_{n=0}^{\infty} D_n$ be the box product of the series $\sum_{n=0}^{\infty} |a_n|$ and $\sum_{n=0}^{\infty} |b_n|$.

Since $\sum_{n=0}^{\infty} |a_n|$ and $\sum_{n=0}^{\infty} |b_n|$ are convergent, $\sum_{n=0}^{\infty} D_n$ is convergent by Theorem 4.9.1. Also, for all n,

$$
\begin{aligned}
|d_n| &= |(a_0+a_1+a_2+\ldots a_n)b_n+a_n(b_0+b_1+b_2+\ldots+b_{n-1})+a_nb_n| \\
&\leqslant (|a_0|+|a_1|+\ldots+|a_{n-1}|)\,|b_n|+|a_n|\,(|b_0|+|b_1|+\ldots+|b_{n-1}|) \\
&\quad +|a_n|\,|b_n| = D_n\,.
\end{aligned}
$$

Therefore, by the comparison test, $\sum_{n=0}^{\infty} d_n$ is absolutely convergent. Therefore, since $\sum_{n=0}^{\infty} c_n$ is a rearrangement of $\sum_{n=0}^{\infty} d_n$, Theorem 4.8.2 gives $\sum_{n=0}^{\infty} c_n = \sum_{n=0}^{\infty} d_n$, and so, by Theorem 4.9.1,

$$\sum_{n=0}^{\infty} c_n = \sum_{n=0}^{\infty} d_n = \left(\sum_{n=0}^{\infty} a_n\right)\left(\sum_{n=0}^{\infty} b_n\right). \quad \square$$

The usefulness of the Cauchy product and of many of the other theorems of this chapter will become apparent in Chapter 7, when we consider power series and the functions which they define.

Exercises 4.9

1. For what values of x is the series $\sum_{n=0}^{\infty} x^n$ convergent and what is its sum to infinity? By forming the Cauchy product of the series with itself deduce the sum to infinity of the series $\sum_{n=0}^{\infty} nx^n$.

2. If $\sum_{n=0}^{\infty} c_n$ is the Cauchy product of the series $\sum_{n=0}^{\infty} (-1)^n/\sqrt{(n+1)}$ with itself, show that $|c_n| \geqslant 1$ for all n. Hence show that the Cauchy product of two conditionally convergent series may be divergent.

Exercises for Chapter 4

1. How many terms of the harmonic series would you take in order to ensure that their sum exceeded 6? How many would you take in order to ensure that their sum exceeded 60? (Do not evaluate the second answer.)

2. Given that $e = \lim_{n\to\infty} (1+1/n)^n$, prove that the series S whose nth term is $(n!/n^n)x^n$ converges if $0 < x < e$ and diverges if $x > e$. Deduce that $\lim n!/(\frac{1}{2}n)^n = 0$.

Given also that $(1+1/n)^n$ is an increasing function of n, for n positive, prove that $n!(e/n)^n$ is an increasing function of n and hence that S diverges if $x = e$. [T.C.]

(It may be assumed that $e > 2$. Also compare with question 4 of exercises for Chapter 3.)

3. Explain the meanings of the following statements:

(a) Σu_n is convergent,

(b) Σu_n is absolutely convergent.

Prove that the series $\sum_{n=1}^{\infty} (-1)^{n+1}(x^n/n)$ is absolutely convergent when $|x| < 1$ and convergent when $x = 1$, stating any tests for convergence which you use.

Show that the series is divergent when $x = -1$. [T.C.]

4. Find all values of x for which each of the following series of real terms is convergent:

(a) $x+\dfrac{x^4}{2}+\dfrac{x^9}{3}+\ldots+\dfrac{x^{n^2}}{n}+\ldots,$

(b) $\dfrac{2x}{1!}+\dfrac{3x^2}{2!}+\ldots+\dfrac{(n+1)x^n}{n!}+\ldots$ [T.C.]

5. In each of the five following pairs of statements, (a) and (b), one statement is true and one statement is false. Write down which is the false statement, and demonstrate its falsity by means of a counter-example.

(i) (a) Every non-empty bounded set of rational numbers has a least upper bound which is a rational number.
(b) Every non-empty bounded set of real numbers has a least upper bound which is a real number.

(ii) $\{s_n\}$ and $\{t_n\}$ are convergent sequences such that

$\lim s_n = S$ and $\lim t_n = T$.

(a) $\lim s_n t_n = ST$.

(b) $\lim s_n/t_n = S/T$.

(iii) (a) If Σu_n is convergent, then $\lim u_n = 0$.

(b) If $\lim u_n = 0$, then Σu_n is convergent.

(iv) (a) If $\lim |u_n/u_{n+1}| = l > 1$, then Σu_n is absolutely convergent.

(b) If $\lim |u_n/u_{n+1}| = 1$, then Σu_n is not convergent.

(v) (a) If Σa_n and Σb_n are series of positive terms, and $\lim_{n \to \infty} a_n/b_n = L$, where L is finite and non-zero, then Σa_n and Σb_n are either both convergent or both divergent.

(b) If Σa_n and Σb_n are series of real terms, and $\lim |a_n/b_n| = L$, where L is finite and non-zero, then Σa_n and Σb_n are either both convergent or both divergent.

6. Prove the following test for the convergence of the alternating series $\Sigma(-1)^n a_n$:

If (1) for all n, $a_n \geqslant a_{n+1} > 0$, and (2) $\lim_{n \to \infty} a_n = 0$, then $\Sigma(-1)^n a_n$ is convergent.

Two of the four series given below are convergent; the other two are not convergent. State, with reasons, which two series are convergent.

$$\sum(-1)^n \frac{n+1}{n}, \quad \sum(-1)^n \frac{n+1}{n^2}$$

$$\sum(-1)^n a_n, \text{ where, } a_n = \begin{cases} \dfrac{1}{n} & (n \text{ even}) \\ \dfrac{1}{n^2} & (n \text{ odd}) \end{cases}$$

$$\sum(-1)^n b_n, \text{ where } b_n = \begin{cases} \dfrac{1}{n^2} & (n \text{ even}) \\ \dfrac{1}{n^3} & (n \text{ odd}) \end{cases}$$

Explain the following statements about the four series:

(a) Of the two convergent series, the convergence of one may be deduced from the alternating series test above; the convergence of the other may not.

(b) Of the two non-convergent series, each satisfies one of the conditions (1) and (2) of the alternating series test, but not the other. [T.C.].

7. State the comparison test for the convergence of series of positive terms. Prove that, if a_n is real and Σa_n is absolutely convergent, then Σa_n^2 is convergent. Show that if Σa_n is convergent but not absolutely convergent, then Σa_n^2 may be divergent.

Prove that $\Sigma\, a_n/(k+a_n)$ (where k is a non-zero constant and $a_n \neq -k$ for all n) is absolutely convergent if and only if Σa_n is absolutely convergent.

Prove that if Σa_n is absolutely convergent, then $\Sigma\, a_n/(a_1+a_2+\ldots+a_n)$ is absolutely convergent provided that all the $s_n = \sum_1^n a_r$ and $s = \lim_{n \to \infty} s_n$ are non-zero. [T.C.]

8. If Σa_n if a convergent series and $a_n > 0$ for all n, prove that $\Sigma a_n a_{n+1}$ also converges.

Show by examples that:

(a) this theorem is not necessarily true if a_n can take any real value;

(b) the converse of this theorem is not necessarily true.

If Σc_n is a convergent series and Σd_n is divergent, prove that $\Sigma(c_n+d_n)$ diverges. Construct a divergent series Σd_n such that $\Sigma(d_n+d_{n+1})$ converges. [T.C.]

9. (i) Assuming that every non-empty set of real numbers that is bounded above has a least upper bound, prove that:

(a) every non-empty set of real numbers that is bounded below has a greatest lower bound, and

(b) every bounded infinite set of real numbers has at least one limit point (accumulation point).

Give examples to show that the greatest lower bound in (a), and the accumulation point in (b), may—but need not—be members of the set concerned.

(ii) Give a formal definition of the limit l of a convergent sequence $\{s_n\}$ of real numbers, in terms of l, N, and ε. If s_n is the sum to n terms of the series

$$1+\tfrac{1}{2}+\tfrac{1}{4}+\ldots+(\tfrac{1}{2})^r+\ldots,$$

determine a suitable N if $\varepsilon = 0.01$. [T.C.]

10. What is meant by saying that a series of real terms Σa_n is "absolutely convergent"? Prove d'Alembert's ratio test for the convergence of a series Σu_n of positive terms.

Explain how d'Alembert's test can be used to investigate the convergence of a series containing both positive and negative terms. State one other test for the convergence of a series of real terms.

Determine the range of convergence of each of the following series:

(a) $\sum_{n=1}^{\infty}(-1)^n n^2 x^n$, (b) $\sum_{n=1}^{\infty}\dfrac{(-1)^n x^n}{2n+1}$. [T.C.]

11. (i) Prove that $\sum_{n=1}^{\infty} 1/n$ is divergent.

(ii) Prove that $\sum_{n=1}^{\infty} \sin^2(x/n)/(n^2-\frac{1}{4})$ is convergent for all real x.

(iii) Find for what real values of x each of the following series is convergent:

$$\sum_{n=1}^{\infty}\frac{(ax)^{2n}}{n},\quad \sum_{n=1}^{\infty}\frac{x^{2n}}{n!}.$$ [B.Ed.]

12. Prove that an absolutely convergent series is convergent.

(i) Give an example of a series which is convergent but not absolutely convergent.

(ii) Find for what real values of x the series $\sum_{n=1}^{\infty} 1/(1+x^{2n})$ is convergent. [B.Ed.]

13. State sufficient conditions to ensure that

$$\left(\sum_{n=0}^{\infty} a_n\right)\left(\sum_{n=0}^{\infty} b_n\right) = \sum_{n=0}^{\infty} c_n,$$

where $c_0 = a_0 b_0$ and $c_n = a_0 b_n + a_1 b_{n-1} + \ldots + a_{n-1} b_1 + a_n b_0$ for $n > 0$.

(i) Obtain the simplest form of c_n for the product when

$$a_n = \frac{x^n}{n!} \quad \text{and} \quad b_n = \frac{(-x)^n}{n!}.$$

(ii) Prove that

$$\left(\sum_{n=1}^{\infty} x^n\right)\left(\sum_{n=1}^{\infty} nx^{n-1}\right) = \sum_{n=1}^{\infty} \tfrac{1}{2}n(n+1)x^n$$

when $|x| < 1$. Hence, or otherwise, obtain the sum of the series

$$\sum_{n=1}^{\infty} n(n+1)x^n \quad \text{when} \quad |x| < 1.$$ [B.Ed.]

14. (a) Determine whether the following series converge or diverge:

(i) $1 - \frac{6}{7} + \frac{8}{10} - \frac{10}{13} + \frac{12}{16} - \frac{14}{19} + \ldots$

(ii) $\frac{1}{3} + \frac{1}{5^2} + \frac{1}{3^3} + \frac{1}{5^4} + \frac{1}{3^5} + \frac{1}{5^6} + \ldots$

(b) Given that $\sum_{n=1}^{\infty} a_n$ is a convergent series of positive terms, prove that $\sum_{n=1}^{\infty} a_n/1+a_n$ and $\sum_{n=1}^{\infty} a_n^3$ are also convergent.

(c) Discuss the convergence, for real values of x, of the series

(i) $\sum_{n=1}^{\infty} \frac{1}{n}\left(\frac{x}{2}\right)^{2n}$. (ii) $\sum_{n=1}^{\infty} \left(x^n + \frac{1}{x^n}\right)$. [B.Ed.]

15. If the sequence a_n of non-negative real numbers is decreasing show that

$$a_{m+1} - a_{m+2} \leqslant a_{m+1} - a_{m+2} + \ldots (-1)^{n-1} a_{m+n} \leqslant a_{m+1},$$

distinguishing between the cases of n odd and n even.

Deduce from the Cauchy principle of convergence that if $\lim a_n = 0$, then $\sum_{1}^{\infty} (-1)^n a_n$ converges. Writing r_m for $\sum_{m+1}^{\infty} (-1)^r a_r$, show that $|r_m| \leqslant a_{m+1}$.

By taking sufficiently many terms to ensure that the error in approximating to $\sum_1^\infty (-1)^r/r^4$ by a finite sum is less than 0.01, show that

$$-0.960 < \sum_1^\infty (-1)^r/r^4 < -0.939. \qquad \text{[B.Ed.]}$$

16. Explain briefly what is meant by a proof by mathematical induction. Prove that, for $m = 1, 2, 3, \ldots,$

$$\sum_{n=1}^{m} \frac{n}{(2n-1)(2n+1)(2n+3)} = \frac{m(m+1)}{2(2m+1)(2m+3)}$$

Discuss for all real x the convergence of the series

$$\sum_{n=1}^{\infty} \frac{nx^n}{(2n-1)(2n+1)(2n+3)}. \qquad \text{[B.Sc.]}$$

17. Σa_n is an infinite series of positive terms; either prove, or disprove by means of a counter-example, the following assertions:

(a) If $\dfrac{a_{n+1}}{a_n} < 1$ for all n, then Σa_n is convergent

(b) If $\dfrac{a_{n+1}}{a_n} > 1$ for all n, then Σa_n is divergent.

Discuss the convergence, for all positive values of x, of the series

$$\Sigma\, n!x^n \quad \text{and} \quad \Sigma n^2 x^{-n} \qquad \text{[B.Sc.]}$$

18. (i) Discuss the convergence of the series

$$\tfrac{1}{1} - \tfrac{2}{3} + \tfrac{3}{5} - \tfrac{4}{7} + \tfrac{5}{9} - \ldots$$

(ii) Find for what real values of x the series

$$\sum \frac{3, 5, \ldots, (2n+1)}{2, 4, \ldots, 2n} (1-x^2)^n$$

(a) diverges, (b) converges absolutely. [B.Sc.]

19. $\sum_1^\infty a_n$ is a convergent series of non-negative terms. If $|b_n| \leqslant K$, a constant, for all n, prove that $\sum_1^\infty a_n b_n$ is convergent.

Prove:

(i) if $\sum_1^\infty a_n x_0^n$ is absolutely convergent then $\sum_1^\infty a_n x^n$ converges absolutely for $|x| \leqslant |x_0|$;

(ii) $\sum_1^\infty 1/n(an+b)^{1/2}$ (a, b positive) is convergent;

(iii) $\sum_1^\infty n!/n^n$ is convergent. [B.Sc.]

20. By considering $\sum_{n+1}^{2n} a_i$, or otherwise, prove that if $\sum_{1}^{\infty} a_n$ is a convergent series of monotonically decreasing positive terms then $na_n \to 0$ as $n \to \infty$.

Prove that:

(i) $\sum_{1}^{\infty} 1/n$ diverges;

(ii) $\sum_{1}^{\infty} (-1)^n/n$ converges. [B.Sc.]

21. If $a_n > 0$, $b_n > 0$ for all n, if Σa_p converges, and if b_n/a_n tends to a finite limit as $n \to \infty$, show that Σb_n converges.

Given that $a_n > 0$ for all n and that Σa_n converges, prove that:

(i) Σa_n^2 converges;

(ii) $\Sigma a_n(a_1 + a_2 + \ldots + a_n)$ converges;

(iii) $\Sigma (\sin a_n)/a_n$ diverges. [B.Sc.]

CHAPTER 5

FUNCTIONS OF A REAL VARIABLE

5.1. Introduction

In Chapter 3 we discussed sequences, i.e. functions whose domain of definition is the set of natural numbers. Sequences are patterns of numbers which change in a succession of jumps.

In this chapter we consider more gradual or continuous change. We shall be concerned with functions whose domain is either the set of real numbers or else some subset of this, e.g. an open or closed interval. The codomain of these functions will also be the set of real numbers. They will be described as functions of a real variable.

DEFINITION. *If a and b are real numbers and $a < b$, then the* open interval (a, b) *is the set* $\{x : a < x < b\}$.

DEFINITION. *If a and b are real numbers and $a < b$, then the* closed interval $[a, b]$ *is the set* $\{x : a \leqslant x \leqslant b\}$.

We also write

$$[a, b) \quad \text{for} \quad \{x : a \leqslant x < b\}$$

$$(a, b] \quad \text{for} \quad \{x : a < x \leqslant b\}$$

$$[a, \infty) \quad \text{for} \quad \{x : a \leqslant x\}$$

and so on.

When the domain and codomain of a function are both the set of real numbers (or some subset), it is natural to picture the domain and codomain as two straight lines. Such a picture is called a graph. We may draw the two lines parallel as in Fig. 5.1 and picture an infinite number of arrows joining points on the domain to points on the codomain; or we may draw the lines perpendicular and obtain the

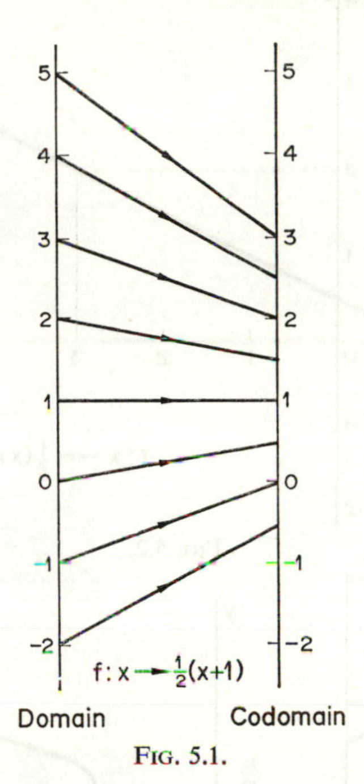

FIG. 5.1.

familiar Cartesian graph as in Fig. 5.2. In the latter case we shall invariably identify the domain with the x-axis and the codomain with the y-axis.

We recall that a function maps every member of the domain on to exactly one member of the codomain. Therefore if we draw a parallel axis graph of a function, each point of the domain has exactly one arrow from it; if we draw a Cartesian graph of a function, then no ordinate (line parallel to the y-axis) may meet the graph of the function more than once. Thus neither of the graphs in Fig. 5.3 represents a function.

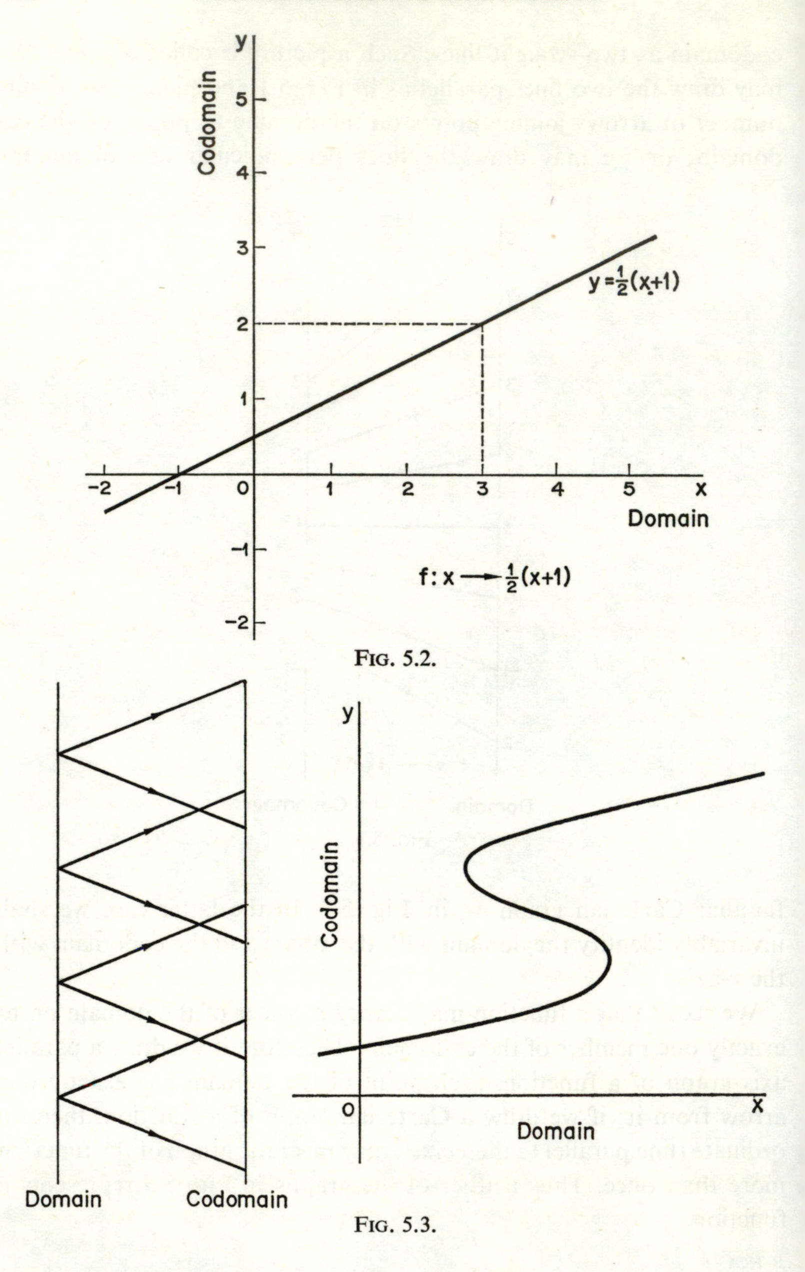
y
Codomain
5
4
3
2
1
0
−1
−2
y = ½(x+1)
−2
−1
1
2
3
4
5
x
Domain
f: x ⟶ ½(x+1)

FIG. 5.2.

Domain
Codomain
y
Codomain
0
x
Domain

FIG. 5.3.

Functions and their domain of definition. The equation

$$f(x) = 3x$$

defines a function f provided that we specify the domain of f. In this case we may choose the whole of the set of real numbers or any subset as the domain of f. The above equation will then define a function of a real variable. So, too, the equation

$$g(x) = x^3 - x$$

defines a function g when we specify the set of real numbers or any of its subsets as the domain of g.

Consider now the equation

$$h(x) = \frac{x-1}{x+2}.$$

If we put $x = -2$ the right-hand side of this equation is meaningless, since division by zero is not possible. So this equation defines a function h, provided that the domain of h does not include the point -2. In the same way the equation

$$j(x) = \sqrt{x}$$

defines a function j provided that the domain of j is some subset of the set of positive real numbers and zero. (We note that we shall always use the symbol $\sqrt{}$ to mean *positive* square root; thus $\sqrt{9}$ is 3 and not -3.)

If we define a function by means of an equation and do not state its domain, then it is to be assumed that the domain of the function is either the set of all real numbers or the set of all those real numbers for which the equation is meaningful. For the sake of brevity we shall sometimes write "the function $f(x) = x^2$" when we mean "the function f defined by the equation $f(x) = x^2$".

Functions defined by more than one equation. Suppose that a stone is dropped over the edge of a cliff. The distance which it has travelled from the starting point after x sec is (approximately) $5x^2$ m. If the cliff is 80 m high, then the stone will come to rest after 4 sec. Thus the function describing the distance of the stone from the starting point for all values of x is given by the equations

$$\begin{cases} f(x) = 0 & \text{if} \quad x < 0, \\ f(x) = 5x^2 & \text{if} \quad 0 \leqslant x \leqslant 4, \\ f(x) = 80 & \text{if} \quad x > 4. \end{cases}$$

This set of equations defines a function whose domain is the set of all real numbers because to every real value of x there is a real value $f(x)$. The Cartesian graph of f is shown in Fig. 5.4.

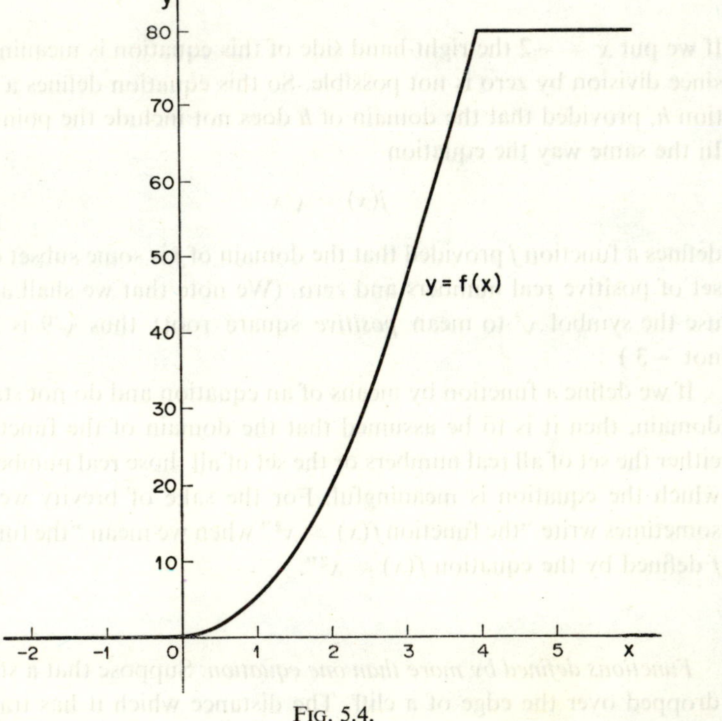

FIG. 5.4.

It is often convenient, both for practical and for theoretical purposes, to define functions by means of more than one equation. Thus we may define the function g by the following equations:

$$g(x) = 3 \quad \text{if} \quad x < 0,$$
$$g(x) = 6 \quad \text{if} \quad x \geqslant 0.$$

The Cartesian graph of g is shown in Fig. 5.5. We see that the function g is discontinuous at $x = 0$ in the sense that, when we draw its Cartesian graph, we have to remove our pencil from the paper when we get

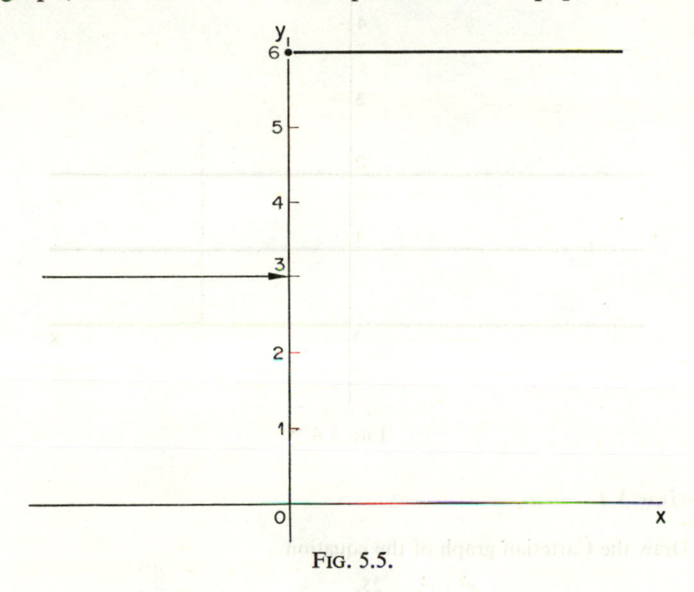

FIG. 5.5.

to $x = 0$ and put it back somewhere else. We discuss the important property of continuity in detail in a later section of this chapter.

Consider now the pathological function h defined by

$$h(x) = 1 \quad \text{if } x \text{ is a rational,}$$
$$h(x) = 2 \quad \text{if } x \text{ is an irrational.}$$

Since both the rationals and the irrationals are dense on the number line, the Cartesian graph of the function h appears to the naked eye as in Fig. 5.6. However, both of the lines representing the function are full of invisible gaps; so, in fact, any ordinate does meet the graph only once, since it will intersect one of the lines and pass through a gap in the other. We shall see later that this function is, in fact, discontinuous everywhere.

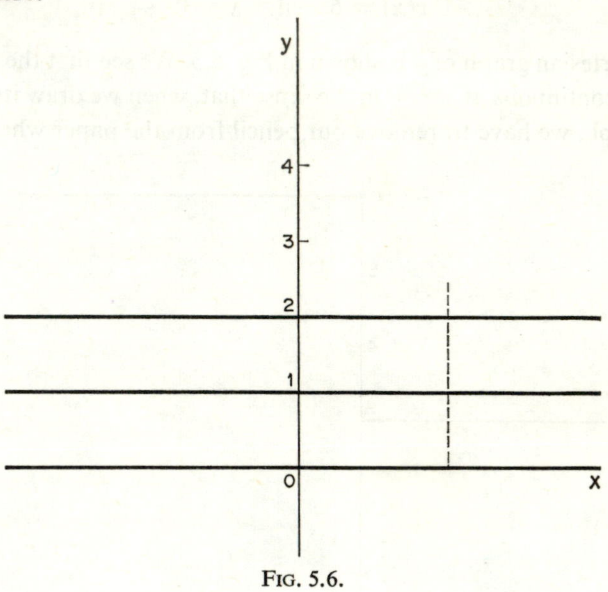

FIG. 5.6.

Exercises 5.1

1. Draw the Cartesian graph of the equation

$$x^2+y^2 = 25.$$

Say why the graph does not represent a function.

2. (i) If $f(x) = |x|$, draw the Cartesian graph of the function f.

(ii) If $g(x) = |x|/x$ state the domain over which g is defined. Draw the Cartesian graph of g.

3. If the function f is defined by the equations

$$\begin{cases} f(x) = x & \text{if} \quad x < 0, \\ f(x) = x^2 & \text{if} \quad 0 \leqslant x \leqslant 1, \\ f(x) = x & \text{if} \quad x > 1. \end{cases}$$

draw the Cartesian graph of f.

5.2. Limits

Monotonic functions. The concept of an increasing or decreasing function is analogous to that of an increasing or decreasing sequence.

DEFINITION. *The function f is an* increasing function *if $f(x) \geqslant f(y)$ whenever $x > y$.*

DEFINITION. *The function f is a* strictly increasing function *if $f(x) > f(y)$ whenever $x > y$.*

DEFINITION. *The function f is a* decreasing function *if $f(x) \leqslant f(y)$ whenever $x > y$.*

DEFINITION. *The function f is a* strictly decreasing function *if $f(x) < f(y)$ whenever $x > y$.*

In the following examples the domain of f is the set of real numbers:

(i) If $f(x) = x^3$, then f is a strictly increasing function.
(ii) If $f(x) = 8$, then f is an increasing function and a decreasing function but neither strictly.
(iii) If $f(x) = 6-8x$, then f is a strictly decreasing function.
(iv) If $f(x) = x^2$, then f is neither an increasing nor a decreasing function.

Bounded functions. We recall that the *range* of the function f, defined over a given domain, is the set $\{f(x): x \text{ belongs to the domain}\}$. If the range of f is a bounded set, then we shall say that the function f is bounded; if the range is a set which is bounded above, then we say that the function f is bounded above, and so on. For example, if we say that the function f is bounded in the closed interval $[a, b]$ we mean that, when the domain of f is $[a, b]$, the range of f is bounded. If we say simply that the function f is bounded we usually mean that it is bounded when the domain of f is the whole of the set of real numbers.

The limit of a function as $x \to \infty$ or as $x \to -\infty$. It is often useful to know what is happening to elements in the range of a function f as we increase without limit elements in the domain. The situation mirrors closely the corresponding situation for sequences, and the relevant definitions are almost identical to the corresponding definitions for sequences.

DEFINITION. *A function $f(x)$ tends to infinity as x tends to infinity (written $f(x) \to \infty$ as $x \to \infty$) if, given* any *real number K, there exists a real number X (which generally depends on the choice of K) such that $f(x) > K$ for all $x \geqslant X$.*

DEFINITION. *A function $f(x)$ tends to minus infinity as x tends to infinity (written $f(x) \to -\infty$ as $x \to \infty$) if, given* any *real number K, there exists a real number X (depending on the choice of K) such that $f(x) < K$ for all $x \geqslant X$.*

DEFINITION. *A function $f(x)$ tends to the limit l as x tends to infinity (written $f(x) \to l$ as $x \to \infty$) if, given* any *real number $\varepsilon > 0$, there exists a real number X (depending on the choice of ε) such that $|f(x) - l| < \varepsilon$ for all $x \geqslant X$.*

We write $\lim_{x \to \infty} f(x) = \infty$, $\lim_{x \to \infty} f(x) = -\infty$, $\lim_{x \to \infty} f(x) = l$ if $f(x)$ tends to infinity, minus infinity and l respectively as x tends to infinity.

In the same way, we may consider what happens to elements in the range of f as we decrease without limit elements in the domain. For the sake of brevity we give only the first definition and leave the reader to formulate the other two.

DEFINITION. *A function $f(x)$ tends to infinity as x tends to minus infinity (written $f(x) \to \infty$ as $x \to -\infty$) if, given* any *real number K, there exists a real number X (depending on the choice of K) such that $f(x) > K$ for all $x \leq X$.*

We write $\lim_{x \to -\infty} f(x) = \infty$, $\lim_{x \to -\infty} f(x) = -\infty$, $\lim_{x \to -\infty} f(x) = l$ if $f(x)$ tends to infinity, minus infinity and l respectively as x tends to minus infinity.

As examples of the consequences of these definitions we have the following two important theorems.

THEOREM 5.2.1. *If $f(x) = x^n$, where n is any natural number, then*

(i) $f(x) \to \infty$ *as* $x \to \infty$,

(ii) $f(x) \to \infty$ *as* $x \to -\infty$ *if n is even,*

(iii) $f(x) \to -\infty$ *as* $x \to -\infty$ *if n is odd.*

Proof. (i) Given any real number K, put $X = \max(K+1, 1)$. For all $x \geq X$,

$$f(x) = x^n \geq X^n \geq X > K.$$

Therefore $f(x) \to \infty$ as $x \to \infty$.

(ii) Given any real number K, put $X = \min(-K-1, -1)$. For all $x \leq X$,

$$f(x) = x^n \geq X^n \geq |X| > K.$$

Therefore $f(x) \to \infty$ as $x \to -\infty$.

(iii) Given any real number K, put $X = \min(K-1, -1)$. For all $x \leq X$,

$$f(x) = x^n \leq X^n \leq X < K.$$

Therefore $f(x) \to -\infty$ as $x \to -\infty$. □

THEOREM 5.2.2. *If $f(x) = x^{1/n}$ for $x \geq 0$, where n is a natural number, then $f(x) \to \infty$ as $x \to \infty$.*

The proof of this theorem is very similar to that of Theorem 3.2.4(vi) and is left to the reader.

We give here one further example. If $f(x) = \sin x$, then $f(x)$ does not have a limit as x tends to infinity or as x tends to minus infinity.

The limit of a function as $x \to a$. We have seen that, for functions of a real variable, we may define limits as x tends to infinity and as x tends to minus infinity which are reminiscent of the limits of sequences. However, with functions of a real variable, we have a further and quite new possibility to consider. We may discuss the behaviour of $f(x)$ as x approaches some finite real number a.

We shall want to say that $f(x)$ tends to l as x tends to a if $f(x)$ approaches l as x approaches a. To put this another way, $f(x)$ will tend to l as x tends to a if we may make $f(x)$ as close as we please to l by taking x sufficiently close to a. So we make the following definition (Fig. 5.7).

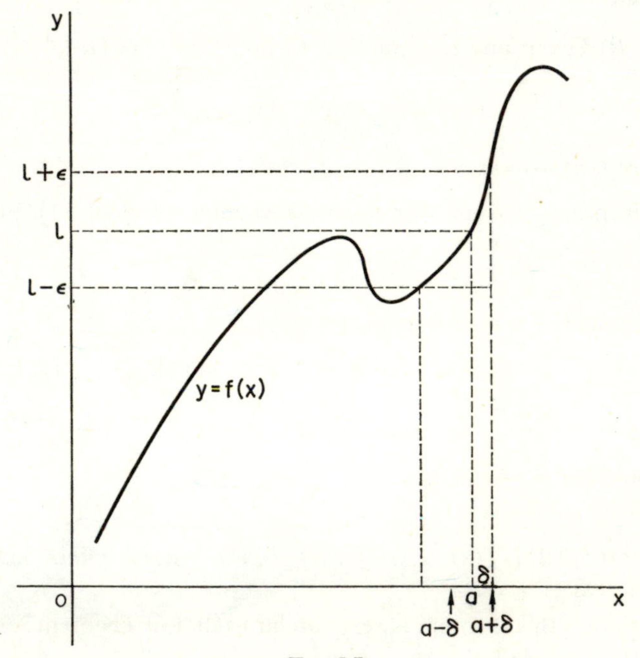

FIG. 5.7.

DEFINITION. *A function $f(x)$ tends to l as x tends to a (written $f(x) \to l$ as $x \to a$) if, given any real number $\varepsilon > 0$, there exists a real number $\delta > 0$ (depending on the choice of ε), such that $|f(x)-l| < \varepsilon$ whenever $0 < |x-a| < \delta$.*

We note particularly that the definition makes no demands at all on the value of $f(a)$.

We also have the following possibilities.

DEFINITION. *A function $f(x)$ tends to infinity as x tends to a (written $f(x) \to \infty$ as $x \to a$) if, given any real number K, there exists a real number $\delta > 0$ (depending on the choice of K), such that $f(x) > K$ whenever $0 < |x-a| < \delta$.*

DEFINITION. *A function $f(x)$ tends to minus infinity as x tends to a (written $f(x) \to -\infty$ as $x \to a$) if, given any real number K, there exists a real number $\delta > 0$ (depending on the choice of K), such that $f(x) < K$ whenever $0 < |x-a| < \delta$.*

We write $\lim f(x) = \infty$, $\lim f(x) = -\infty$, $\lim f(x) = l$ if $f(x)$ tends to infinity, minus infinity and l respectively as x tends to a.

We illustrate these definitions by means of the following examples. In order to obtain a clearer picture of what is happening the reader is advised to sketch a Cartesian graph of each function.

(i) If $f(x) = 1/x^2$:

(a) $f(x) \to \infty$ as $x \to 0$,

(b) $f(x) \to \frac{1}{4}$ as $x \to 2$,

(c) $f(x) \to 0$ as $x \to -\infty$.

(ii) If $f(x) = 1/\{(x-1)(x+2)^4\}$:

(a) $f(x) \to -\infty$ as $x \to -2$,

(b) $f(x)$ does not tend to a limit as $x \to 1$. Why not?

(iii) If $f(x) = (x^2-1)/(x-1)$, then $f(1)$ is not defined, since division by zero is not possible. We prove that $f(x) \to 2$ as $x \to 1$. If $x \neq 1$, then $f(x) = x+1$. Given $\varepsilon > 0$, put $\delta = \varepsilon$. Then $|f(x)-2| = |x+1-2| = |x-1| < \varepsilon$ whenever $0 < |x-1| < \delta$. Therefore $f(x) \to 2$ as $x \to 1$. We stress once more that, in order to evaluate the limit of $f(x)$ as x tends to 1, we do not need to know the value of $f(1)$.

(iv) If f is defined by

$$\begin{cases} f(x) = x^2 & \text{if} \quad x \geqslant 1, \\ f(x) = x & \text{if} \quad x < 1 \end{cases}$$

we shall prove that $f(x) \to 1$ as $x \to 1$.

If $x > 1$, $\qquad f(x)-1 = x^2-1 = (x+1)(x-1)$.

If $x < 1$, $\qquad |f(x)-1| = 1-x$.

Given $\varepsilon > 0$, put $\delta = \min(\varepsilon/2, 1)$. Then

$$\text{if } x > 1, \; |f(x)-1| = (x-1)(x+1) < (\varepsilon/2)\cdot 2 = \varepsilon \quad \text{whenever } x-1 < \delta,$$

$$\text{if } x < 1, \; |f(x)-1| = 1-x < \varepsilon \text{ whenever } 1-x < \delta.$$

Therefore $f(x) \to 1$ as $x \to 1$.

(v) If f is defined by

$$\begin{cases} f(x) = 10 & \text{if} \quad x \geqslant 0, \\ f(x) = 5 & \text{if} \quad x < 0, \end{cases}$$

(a) $f(x) \to 10$ as $x \to 7$,

(b) $f(x) \to 5$ as $x \to -3$,

(c) $f(x)$ does not tend to a limit as $x \to 0$.

(vi) If f is defined by

$$\begin{cases} f(x) = x & \text{if } x \text{ is a rational,} \\ f(x) = 0 & \text{if } x \text{ is an irrational,} \end{cases}$$

we shall prove that $f(x) \to 0$ as $x \to 0$.

Given $\varepsilon > 0$, put $\delta = \varepsilon$.

If x is a rational $|f(x)-0| = |x| < \varepsilon$ whenever $0 < |x-0| < \delta$.

If x is an irrational $|f(x)-0| = 0 < \varepsilon$ whenever $0 < |x-0| < \delta$

Therefore $f(x) \to 0$ as $x \to 0$.

$f(x)$ does not tend to a limit as $x \to a$ for any non-zero value of a. Why not?

We now prove two theorems which will enable us to write down immediately the limits of functions involving powers of x.

THEOREM 5.2.3. *If n is a natural number and if a is any real number, then*

$$x^n \to a^n \text{ as } x \to a.$$

Proof. $x^n - a^n = (x-a)(x^{n-1}+x^{n-2}a+x^{n-3}a^2+\ldots+xa^{n-2}+a^{n-1})$

If $|x-a| < 1$, $|x^{n-1}+x^{n-2}a+x^{n-3}a^2+\ldots+xa^{n-2}+a^{n-1}|$

$$\leqslant |x|^{n-1}+|x|^{n-2}|a|+|x|^{n-3}|a|^2+\ldots+|x|\,|a|^{n-2}+|a|^{n-1}$$
$$< n(|a|+1)^{n-1} = M, \text{ say.}$$

Given $\varepsilon > 0$, put $\delta = \min(1, \varepsilon/M)$. Then, whenever $0 < |x-a| < \delta$,

$$|x^n-a^n| = |x-a|\,|x^{n-1}+x^{n-2}a+x^{n-3}a^2+\ldots+xa^{n-2}+a^{n-1}|$$
$$< \delta M \leqslant \varepsilon.$$

Therefore $x^n \to a^n$ as $x \to a$. □

THEOREM 5.2.4. *If n is a natural number and if a is a real number greater than zero, then*

$$x^{1/n} \to a^{1/n} \quad \text{as} \quad x \to a.$$

(We note that $n^{1/n}$ means the positive nth root of x.)

Proof. Let $x > 0$. Put $y = x^{1/n}$ and $b = a^{1/n}$.

$$x-a = y^n-b^n = (y-b)(y^{n-1}+y^{n-2}b+\ldots+b^{n-1}).$$

Therefore $$|y-b| = \frac{|y^n-b^n|}{y^{n-1}+\ldots+b^{n-1}} < \frac{|y^n-b^n|}{b^{n-1}}.$$

Given $\varepsilon > 0$, put $\delta = b^{n-1}\varepsilon$. Then, whenever $0 < |x-a| < \delta$,

$$|x^{1/n}-a^{1/n}| = |y-b| < \frac{|x-a|}{b^{n-1}} < \frac{\delta}{b^{n-1}} = \varepsilon.$$

Therefore $x^{1/n} \to a^{1/n}$ as $x \to a$. □

Trigonometric functions. We shall discuss trigonometric functions in detail in Chapter 7. However, since the reader is familiar with these functions we shall wish to use them in order to illustrate some of the concepts of this and the following chapters. Consequently we give here proofs of some of the properties of trigonometric functions which we want to use. The proof of the following theorem must be regarded as provisional because we break our rule and use geometrical intuition in order to prove the first part.

THEOREM 5.2.5. (i) *If* $-\pi/2 < x < \pi/2$, *then* $|\sin x| \leqslant |x| \leqslant |\tan x|$.

(ii) *If* $-\pi/2 < x < \pi/2$, *then* $1-x^2 \leqslant \cos^2 x \leqslant 1$.

(iii) $\lim\limits_{x\to 0} \cos x = 1$.

(iv) $\lim\limits_{x\to 0} (\sin x)/x = 1$.

(v) $\lim\limits_{x\to 0} (\tan x)/x = 1$.

Proof. (i) Suppose first that x is acute. We draw a circle of radius r with centre O and mark a sector AOB whose angle is x (Fig. 5.8). AC is a tangent to the circle.

Area of triangle AOC > area of sector AOB > area of triangle AOB. Therefore

$$\tfrac{1}{2}r^2 \tan x > \tfrac{1}{2}r^2 x > \tfrac{1}{2}r^2 \sin x,$$

$$\tan x > x > \sin x.$$

The result (i) follows in view of the fact that $\sin(-x) = -\sin x$ and $\tan(-x) = -\tan x$.

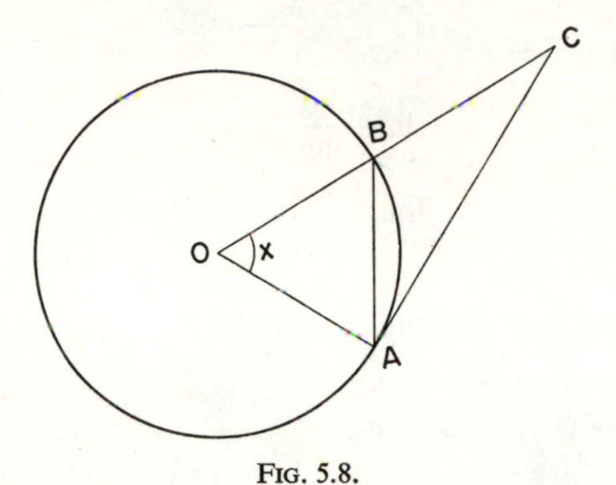

FIG. 5.8.

(ii) We know that $\cos x \leqslant 1$. Also since $\cos^2 x = 1-\sin^2 x$, result (i) gives

$$1-x^2 \leqslant 1-\sin^2 x = \cos^2 x \leqslant 1.$$

(iii) Given $\varepsilon > 0$, put $\delta = \min(1, \varepsilon)$. Then $\delta^2 \leqslant \varepsilon^2 \leqslant \varepsilon$. Therefore, whenever $0 < |x| < \delta$,

$$|1-\cos x| < |1-\cos^2 x| < x^2 < \delta^2 \leqslant \varepsilon.$$

Therefore $\lim\limits_{x\to 0} \cos x = 1$.

(iv) If $-\pi/2 < x < \pi/2$, then (i) gives

$$1 \leqslant \frac{x}{\sin x} \leqslant \frac{1}{\cos x} \quad \text{provided } x \neq 0.$$

Thus
$$1 \geqslant \frac{\sin x}{x} \geqslant \cos x.$$

But $\lim\limits_{x\to 0} \cos x = 1$.

Therefore
$$\lim_{x\to 0} \frac{\sin x}{x} = 1.$$

(v) If $-\pi/2 < x < \pi/2$ then (i) gives

$$\cos x < \frac{x}{\tan x} < 1 \text{ provided } x \neq 0.$$

But $\lim_{x \to 0} \cos x = 1$.

Therefore $$\lim_{x \to 0} \frac{x}{\tan x} = 1.$$

Thus $$\lim_{x \to 0} \frac{\tan x}{x} = 1. \square$$

THEOREM 5.2.6. *If a is any real number:*

(i) $\sin x \to \sin a$ as $x \to a$.

(ii) $\cos x \to \cos a$ as $x \to a$.

Proof. (i) $|\sin x - \sin a| = \left|2 \cos\left(\frac{x+a}{2}\right) \sin\left(\frac{x-a}{2}\right)\right|$

$\leqslant 2\left|\cos\left(\frac{x+a}{2}\right)\right| \left|\sin\left(\frac{x-a}{2}\right)\right| \leqslant 2\left|\sin\left(\frac{x-a}{2}\right)\right| \leqslant |x-a|.$

Given $\varepsilon > 0$, put $\delta = \varepsilon$. Then

$$|\sin x - \sin a| < \varepsilon \text{ whenever } 0 < |x-a| < \delta.$$

Therefore $\sin x \to \sin a$ as $x \to a$.

(ii) $|\cos x - \cos a| = \left|-2 \sin\left(\frac{x+a}{2}\right) \sin\left(\frac{x-a}{2}\right)\right| \leqslant 2\left|\sin\left(\frac{x+a}{2}\right)\right|$

$\left|\sin\left(\frac{x-a}{2}\right)\right| \leqslant 2|\sin(x-a/2)| \leqslant |x-a|.$

Given $\varepsilon > 0$, put $\delta = \varepsilon$. Then

$$|\cos x - \cos a| < \varepsilon \text{ whenever } 0 < |x-a| < \delta. \square$$

Exercises 5.2

1. Prove that, if a function is monotonic in the closed interval $[a, b]$, then it is bounded in $[a, b]$.

2. In each of the following cases state the limit of the function $f(x)$, if it exists, as $x \to 3$:

(i) $f(x) = x^2$.

(ii) $f(x) = \dfrac{1}{(x-3)^4}$.

(iii) $f(x) = \dfrac{1}{(x-3)^3}$

(iv) $f(x) = \dfrac{1}{(x-3)^2(x-5)}$.

(v) f is defined by:

$$\begin{cases} f(x) = 6 & \text{if} \quad x \geqslant 3, \\ f(x) = 2x & \text{if} \quad x < 3. \end{cases}$$

(vi) $f(x) = [x]$, where $[x]$ means the integral part of x; thus, for example, $f(7) = 7, f(2\frac{3}{4}) = 2$, and $f(-3\frac{1}{2}) = -4$.

(vii) f is defined by:

$$\begin{cases} f(x) = 6 & \text{if } x \text{ is a rational,} \\ f(x) = 3 & \text{if } x \text{ is an irrational.} \end{cases}$$

3. If $f(x) = x^2$, prove *from the definition* that $f(x) \to 9$ as $x \to 3$.

4. If $f(x) = |x|$, prove that $f(x) \to 0$ as $x \to 0$.

5. State the following limits if they exist:

(i) $\lim_{x \to 0} \left(\frac{x}{\sin x}\right)$.

(ii) $\lim_{x \to 0} \left(\frac{\sin^2 x}{x}\right)$

(iii) $\lim_{x \to \infty} \sin x$.

(iv) $\lim_{x \to \infty} \left(\frac{\sin x}{x}\right)$.

(v) $\lim_{x \to 0} \left(\frac{\cos x}{x}\right)$

(vi) $\lim_{x \to \infty} \cos\left(\frac{1}{x}\right)$.

(vii) $\lim_{x \to 0} \cos\left(\frac{1}{x}\right)$.

5.3. Properties of limits

In section 3.3 we proved a number of theorems about the limits of sequences. There are theorems about functions of a real variable corresponding to all these theorems except Theorem 3.3.2. We shall not prove or even state all these theorems although we may use their implications in the subsequent theory. However, we shall state the most important theorem and prove part of it.

THEOREM 5.3.1. *f and g are functions of a real variable. h, j, k, and F are all functions, defined by the following equations:*

$$h(x) = f(x)+g(x), \quad j(x) = f(x)+g(x), \quad k(x) = f(x)\,g(x)$$

$$F(x) = \frac{1}{f(x)}.$$

(A) *If* $f(x) \to l$ *as* $x \to \infty$, $g(x) \to m$ *as* $x \to \infty$, *then*

(i) $h(x) \to l+m$ *as* $x \to \infty$,

(ii) $j(x) \to l-m$ *as* $x \to \infty$,

(iii) $k(x) \to lm$ *as* $x \to \infty$,

(iv) $F(x) \to 1/l$ *as* $x \to \infty$ *provided* $l \neq 0$.

(B) *If* $f(x) \to l$ *as* $x \to -\infty$, $g(x) \to m$ *as* $x \to -\infty$, *then*

(i) $h(x) \to l+m$ *as* $x \to -\infty$,

(ii) $j(x) \to l-m$ *as* $x \to -\infty$,

(iii) $k(x) \to lm$ *as* $x \to -\infty$,

(iv) $F(x) \to 1/l$ *as* $x \to -\infty$ *provided* $l \neq 0$.

(C) *If* $f(x) \to l$ *as* $x \to a$, $g(x) \to m$ *as* $x \to a$, *then*

(i) $h(x) \to l+m$ *as* $x \to a$,

(ii) $j(x) \to l-m$ *as* $x \to a$,

(iii) $k(x) \to lm$ *as* $x \to a$,

(iv) $F(x) \to 1/l$ *as* $x \to a$ *provided* $l \neq 0$.

Proof. To illustrate the method of adapting the proof of Theorem 3.3.3, we prove one part of this theorem, C(iii).

$f(x) \to l$ as $x \to a$. Therefore we can find a real number $\delta_0 > 0$ such that $|f(x)| < |l|+1$ whenever $0 < |x-a| < \delta_0$.

$$|k(x)-lm| = |f(x)\, g(x)-lm| = |f(x)\,[g(x)-m]+m[f(x)-l]|.$$

Given $\varepsilon > 0$, there is a real number $\delta_1 > 0$ such that

$$|f(x)-l| < \frac{\varepsilon}{2|m|+1} \quad \text{whenever} \quad 0 < |x-a| < \delta_1$$

and there is a real number $\delta_2 > 0$ such that

$$|g(x)-m| < \frac{\varepsilon}{2|l|+2} \quad \text{whenever} \quad 0 < |x-a| < \delta_2.$$

Put $\delta = \min(\delta_0, \delta_1, \delta_2)$. Then, for all x satisfying $0 < |x-a| < \delta$

$$|k(x)-lm| = |f(x)[g(x)-m]+m[f(x)-l]| \leqslant |f(x)|\,|g(x)-m| + |m|\,|f(x)-l| < \frac{(|l|+1)\varepsilon}{2|l|+2} + \frac{|m|\,\varepsilon}{2|m|+1} < \varepsilon$$

Therefore $k(x) \to lm$ as $x \to a$. □

Corollary. *If* $f(x) \to l$ *as* $x \to a$ *and* $g(x) \to m$ *as* $x \to a$ *and if* $m \neq 0$, *then* $f(x)/g(x) \to l/m$ *as* $x \to a$.

Exercises 5.3

1. Prove Theorem 5.3.1, parts B(iii) and C(iv).
2. What is the limit of $f(x)$ as $x \to 0$ in each of the following cases?
   (i) $f(x) = \sin x + \cos x$.
   (ii) $f(x) = (x+1)^2 \left[\cos x + \frac{x(x-2)^3}{\tan x}\right]$.

5.4. Continuity

In the introductory chapter we mentioned the use of graphs of functions in the solution of equations. We suggested that if we had to find the number of the real roots of the equation

$$x^2 = \cos x$$

we might draw graphs of $y = x^2$ and $y = \cos x$ and look for their points of intersection (see Fig. I.4). In concluding from the graph that this particular equation has two real roots, we make the assumption that the curves representing the functions are continuous.

We think intuitively of a curve as continuous if we are able to draw it without lifting the pencil from the paper; thus the curve of Fig. 5.9a is continuous whereas the curve of Fig. 5.9b is not. We want to obtain,

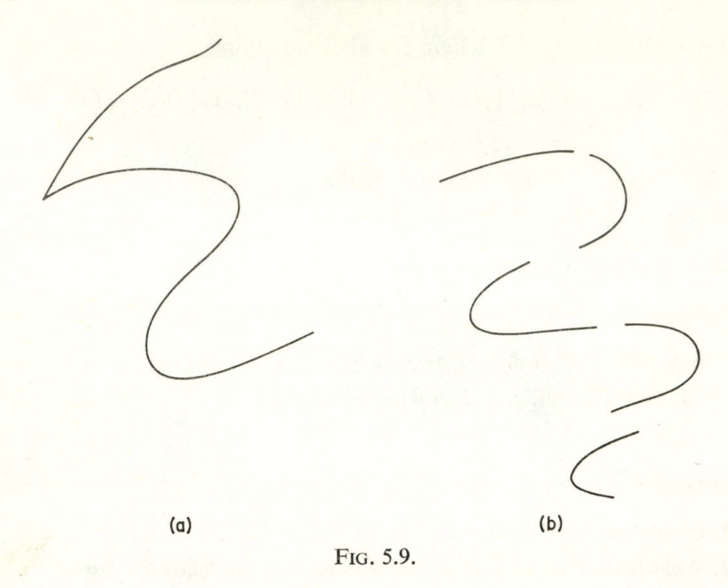

FIG. 5.9.

from this intuitive notion of continuity, a definition in the language of analysis in order that we may ascertain whether various familiar functions are really continuous and in order that we may prove that continuous functions have certain intuitively obvious properties.

So that we are able to give an analytic definition of continuity, we reformulate our intuitive notion of continuity.

Suppose that P is a point on the Cartesian graph of the function f [say P is the point $(a, f(a))$]. Then, if the graph is continuous, the graph arcs to either side of P must meet up at P, that is $f(x)$ must tend to $f(a)$ as x tends to a (compare situations (a) and (b) in Fig. 5.10). So we make the following definition of continuity.

DEFINITION. *We shall say that the function f is continuous at the point a if $f(x) \to f(a)$ as $x \to a$.*

We note that we have defined continuity *at a point* (a strange concept intuitively). To bring our definitions more into line with intuition we shall use the following terminology.

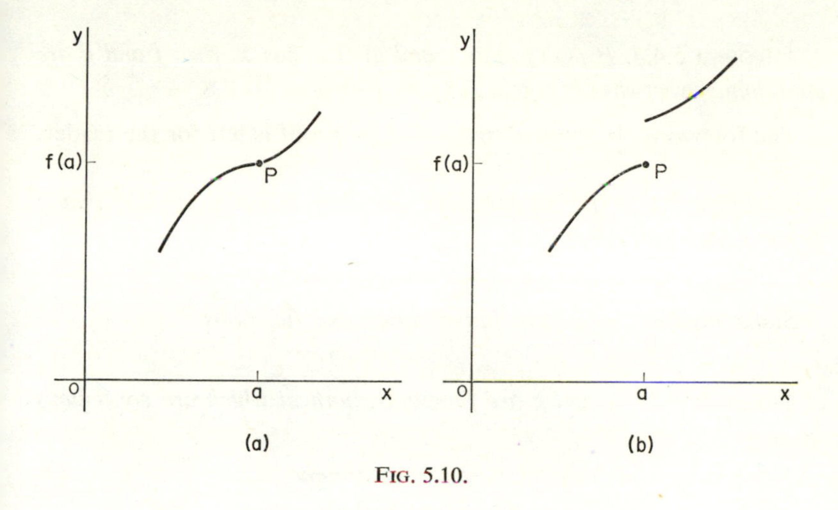

FIG. 5.10.

DEFINITION. *A function f is continuous in an open interval if it is continuous at every point of that interval.*

DEFINITION. *A function f is continuous everywhere if it is continuous at a for all real values of a.*

Whereas intuitively we think of continuity as the property of a function throughout some interval, analytically we define continuity first at a point and only subsequently over an interval.

Examples of continuous functions. As a direct consequence of Theorems 5.2.3, 5.2.4, and 5.2.6, we have the following theorems.

THEOREM 5.4.1. *If n is a natural number and if $f(x) = x^n$, then f is continuous everywhere.*

THEOREM 5.4.2. *If n is a natural number and if $f(x) = x^{1/n}$, then f is continuous for all $x > 0$.*

THEOREM 5.4.3. *If* $f(x) = \sin x$ *and* $g(x) = \cos x$, *then* f *and* g *are continuous everywhere.*

The following theorem is trivial and its proof is left for the reader.

THEOREM 5.4.4. *If* c *is any real constant and if* $f(x) = c$, *then* f *is continuous everywhere.*

Sums, products, and quotients of continuous functions

THEOREM 5.4.5. f *and* g *are functions, both of which are continuous at* a.

(i) *If* $h(x) = f(x)+g(x)$, *then* h *is continuous at* a.
(ii) *If* $j(x) = f(x)-g(x)$, *then* j *is continuous at* a.
(iii) *If* $k(x) = f(x)\,g(x)$, *then* k *is continuous at* a.
(iv) *If* $F(x) = 1/f(x)$, *then* F *is continuous at* a *provided* $f(a) \neq 0$.

Proof. The theorem follows immediately from Theorem 5.3.1(C). For example, we prove (iii).

Since f and g are continuous at a, $f(x) \to f(a)$ as $x \to a$ and $g(x) \to g(a)$ as $x \to a$. Therefore, by Theorem 5.3.1(C) (iii)

$$f(x)\,g(x) \to f(a)\,g(a) \quad \text{as} \quad x \to a.$$

Thus $k(x) \to k(a)$ as $x \to a$ and so k is continuous at a. □

From Theorem 5.4.5 we are able to deduce that all polynomials are continuous and all rational functions are continuous except at points where they are undefined (because the denominator is zero). For example,

$$f(x) = x^3+3x^2-2x+7,$$

$$f(x) = \frac{x-1}{x-3},$$

$$f(x) = \frac{(x+2)^2\,(x-1)}{(x^4+4x^2+2)\,(x+5)},$$

are all continuous at all those points for which they are defined.

Composite functions (function of a function). Consider the function f defined by

$$f(x) = \sqrt{(x^2+x+1)}.$$

We may think of such a function as built up in two stages. First we compute x^2+x+1; then we take the square root. So f is built from the functions

$$g(x) = x^2+x+1 \quad \text{and} \quad h(x) = \sqrt{x}$$

and f is the composite function, given by $f = h \circ g$ (see p. 9).

The reason for breaking functions down in this way is that it makes their continuity and differentiability properties easier to prove and their derivatives easier to compute.

THEOREM 5.4.6. *If $f = h \circ g$, where g is continuous at a, h is continuous at b and $b = g(a)$ (Fig. 5.11), then f is continuous at a.*

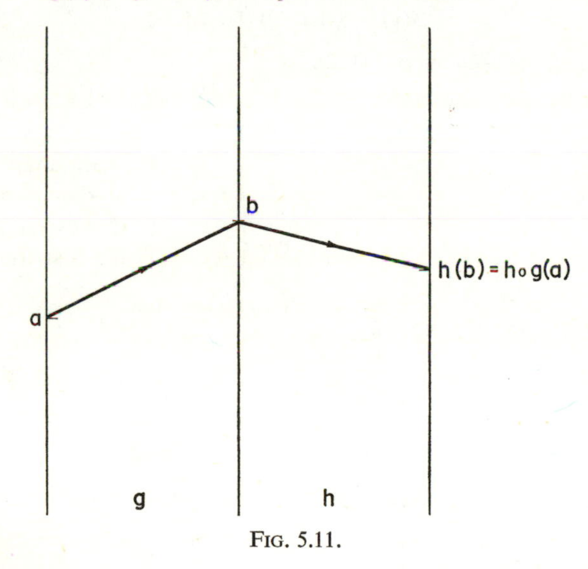

FIG. 5.11.

Proof. We have to show that $f(x) \to f(a)$ as $x \to a$. Put $g(x) = u$. Then $f(x) = h(u)$. Also $f(a) = h(b)$. Since h is continuous at b, given

$\varepsilon > 0$, there is a real number $\delta_1 > 0$ such that

$$|f(x)-f(a)| = |h(u)-h(b)| < \varepsilon \quad \text{whenever} \quad |u-b| < \delta_1.$$

Since g is continuous at a, given $\delta_1 > 0$, there is a real number $\delta_2 > 0$ such that

$$|u-b| = |g(x)-g(a)| < \delta_1 \quad \text{whenever} \quad |x-a| < \delta_2.$$

Thus, whenever $|x-a| < \delta_2$, $|f(x)-f(a)| < \varepsilon$. Therefore $f(x) \to f(a)$ as $x \to a$. □

Thus if $f(x) = \sqrt{(x^2+x+1)}$, f is continuous whenever it is defined, since, by Theorems 5.4.1, 5.4.2, and 5.4.5, g and h are continuous.

Exercises 5.4

1. Use Theorem 5.3.1 to prove Theorem 5.4.5. (iv).

2. Explain why the function f defined by

$$f(x) = \sqrt{[(x-2)(1-x)]}$$

is continuous in the open interval (1, 2).

3. Sketch the Cartesian graphs of the following functions; in each case say where f is continuous and where it is discontinuous:

(i) $f(x) = x^3+7$.

(ii) $f(x) = \dfrac{x+3}{x+5}$.

(iii) $f(x) = \sqrt{\left(\dfrac{x^2+1}{x^2-1}\right)}$.

(iv) $\begin{cases} f(x) = x+3 & \text{if } |x| \leqslant 2 \\ f(x) = x^2 & \text{if } |x| > 2 \end{cases}$

(v) $\begin{cases} f(x) = 0 & \text{if } x \text{ is a rational,} \\ f(x) = x & \text{if } x \text{ is an irrational.} \end{cases}$

(vi) $\begin{cases} f(x) = 1 & \text{if } x \text{ is a rational,} \\ f(x) = 3 & \text{if } x \text{ is an irrational.} \end{cases}$

(vii) $f(x) = \lim\limits_{n \to \infty} \dfrac{x^n}{x^n+1}$.

4. If f is defined by the equations

$$\begin{cases} f(x) = \cos\dfrac{1}{x} & \text{if } x \neq 0, \\ f(x) = 0 & \text{if } x = 0. \end{cases}$$

sketch the Cartesian graph of f. Show that f is discontinuous at zero. Explain why it is continuous everywhere else.

5. Show, by giving examples, that all the following may occur:

(i) f is continuous everywhere.

(ii) f is continuous nowhere.

(iii) f is continuous at only one point.

(iv) f is continuous at exactly four points.

(v) f is continuous everywhere except at two points.

(vi) f is continuous at an infinite number of points and discontinuous at an infinite number of points.

5.5. The place of pathological functions in real analysis

Commonly occurring functions are invariably continuous except sometimes at a few isolated points. Functions, useful in physics and elsewhere, formed by sticking two or more pieces of familiar functions together at their ends, are also readily checked for continuity or discontinuity at the join points. So why do we consider pathological functions like those given in exercises 5.4, question 3, (v) and (vi), and why do we make so much of continuity?

The answer is this. Experience with common functions may lead us to wonder whether certain propositions are true. We may, in the light of our experience with familiar functions, believe that every function is continuous at least somewhere or that, if a function is continuous at a point, then it is continuous in some interval containing that point. Both these statements are fallacious, as the examples mentioned above show. So pathological functions are used as counter examples to falsify statements which we might otherwise believe to be true. This, in turn, helps us to see exactly what conditions it is necessary to impose on a function in order that a particular theorem about it may be valid.

Exercises 5.5

By considering functions which you have already met in this chapter, and others if necessary, try to decide whether each of the following statements is generally true or not. If it is not always true then give a counter example.

1. If f is an increasing function, then it is continuous everywhere.

2. If f tends to a finite limit, both as x tends to infinity and as x tends to minus infinity, then f is bounded.

3. If f is continuous and bounded, then f tends to a finite limit as x tends to infinity.

4. If f and g are both discontinuous at the point 2, then so is h, where h is defined by $h(x) = f(x)+g(x)$.

5. If f and g are both discontinuous at the point 2, then so is h, where h is defined by $h = g \circ f$,

6. If a bounded, continuous function tends to zero as x tends to infinity and as x tends to minus infinity, then it is constant.

5.6. The nature of discontinuities

One-sidedness. Consider the function f defined by

$$\begin{cases} f(x) = x^2 & \text{if} \quad x \geqslant 2, \\ f(x) = x & \text{if} \quad x < 2. \end{cases}$$

f is continuous everywhere except at 2. If x approaches 2 from the right, then $f(x)$ approaches 4; if x approaches 2 from the left, then $f(x)$ approaches 2 (again the reader is recommended to sketch the graph of f). In order to describe this situation formally we make the following definitions.

DEFINITION. *We shall say that $f(x)$ tends to l as x tends to a from above (written $f(x) \to l$ as $x \to a+0$ or $\lim\limits_{x \to a+} f(x) = l$) if, given $\varepsilon > 0$, we can find a real number $\delta > 0$, such that*

$$|f(x)-l| < \varepsilon \quad \textit{whenever} \quad 0 < x-a < \delta.$$

DEFINITION. *We shall say that $f(x)$ tends to l as x tends to a from below (written $f(x) \to l$ as $x \to a-0$ or $\lim\limits_{x \to a-} f(x) = l$) if, given $\varepsilon > 0$, we can find a real number $\delta > 0$, such that*

$$|f(x)-l| < \varepsilon \quad \textit{whenever} \quad 0 < a-x < \delta.$$

DEFINITION. *f is said to be continuous on the right at a if* $f(x) \to f(a)$ *as* $x \to a+0$.

DEFINITION. *f is said to be continuous on the left at a if* $f(x) \to f(a)$ *as* $x \to a-0$.

For the example given above, $\lim_{x\to 2+} f(x) = 4$, $\lim_{x\to 2-} f(x) = 2$, and, since $f(2) = 4$, f is continuous on the right at 2 but it is not continuous on the left.

Or again, if $f(x) = [x]$ (see exercises 5.2, question 2(vi) for the definition of $[x]$), then f is continuous everywhere except at the integers, f is continuous on the right at the integers, and $\lim_{x\to n+} f(x) = n$, $\lim_{x\to n-} f(x) = n-1$.

THEOREM 5.6.1. $\lim_{x\to a+} f(x) = l$ *and* $\lim_{x\to a-} f(x) = l$ *if and only if* $\lim_{x\to a} f(x) = l$

The proof follows immediately from the definitions.

THEOREM 5.6.2. *f is continuous at a if and only if f is continuous on the right at a and f is continuous on the left at a.*

The proof of this theorem follows immediately from Theorem 5.6.1.

Continuity in a closed interval. Whenever we require a function to be continuous in a closed interval $[a, b]$ we are not interested in its behaviour outside $[a, b]$ (we do not even demand that it should be defined outside $[a, b]$). Therefore we do not require of the function complete continuity at the points a and b.

DEFINITION. *A function is said to be continuous in the closed interval* $[a, b]$ *if it is continuous at all points of the open interval* (a, b) *and is also continuous on the right at a and continuous on the left at b.*

Removable and non-removable discontinuities. Sometimes an equation defines a function everywhere except at one or more isolated points. For example, the function f given by

$$f(x) = \frac{x^2-1}{x-1}$$

is defined everywhere except at the point 1. In fact, when $x \neq 1$, $f(x) = x+1$, so that f is continuous wherever it is defined. Now we may complete the definition of f for all real values of x by defining $f(1)$ to be anything we please. However, since $f(x) \to 2$ as $x \to 1$, there is something natural about completing the definition of f by making $f(1) = 2$. In this way we make f continuous for all values of x.

DEFINITION. *f is a function which is continuous to either side of the point a but is discontinuous at a. It may or may not be defined at a. If, by defining or redefining the value of $f(a)$, we may make f continuous at a, then we shall say that f has a* removable discontinuity *at a. Otherwise we shall say that the discontinuity at a is* non-removable.

For the example given above f has a removable discontinuity at 1. We remove it by defining $f(1) = 2$.

THEOREM 5.6.3. *f is a function which is continuous to either side of the the point a but which is discontinuous at a. Then f has a removable discontinuity at a if and only if $f(x)$ tends to a finite limit as x tends to a.*

Proof. First suppose that f tends to a finite limit as x tends to a. Call this limit l. Define or redefine $f(a)$ by $f(a) = l$. Then $f(x) \to f(a)$ as $x \to a$, and so f is continuous at a.

Now suppose that f has a removable discontinuity at a. Then we may redefine $f(a)$ so that f is continuous at a, or, in other words, so that $f(x) \to f(a)$ as $x \to a$. Evidently this is only possible if $f(x)$ tends to a finite limit as x tends to a. □

For example, if $f(x) = 1/x$, then f has a non-removable discontinuity at zero since $f(x)$ has no finite limit as x tends to zero.

Or consider $f(x) = \cos 1/x$. The reader should be able to see from a sketch graph that $f(x)$ does not tend to a limit as x tends to zero. So f has a non-removable discontinuity at zero.

Jump discontinuities. There is one particularly simple type of non-removable discontinuity—the jump discontinuity. This arises most characteristically when we have a function defined in terms of two equations and the two pieces of curve fail to meet at the transition point from one equation to the other. Thus the function f defined by

$$\begin{cases} f(x) = x^2 & \text{if} \quad x \geqslant 3, \\ f(x) = x & \text{if} \quad x < 3, \end{cases}$$

has a jump discontinuity at 3.

DEFINITION. *If f is continuous on either side of a and if* $\lim_{x \to a+} f(x) = l$, $\lim_{x \to a-} f(x) = L$, *where $l \neq L$, then f is said to have a* jump discontinuity *at the point a.*

A theorem about limits

THEOREM 5.6.4. *If g is a function continuous at l and if $f(x)$ tends to l as x tends to a, then $g(f(x))$ tends to $g(l)$ as x tends to a.*

Proof. If f is discontinuous at a then its discontinuity is removable by Theorem 5.6.3. Remove the discontinuity by defining $f(a) = l$. Then f is continuous at a and so, by Theorem 5.4.6, $g \circ f$ is continuous at a. Therefore $g(f(x))$ tends to $g[f(a)] = g(l)$ as x tends to a.

Exercises 5.6

1. Show that if f is defined everywhere except at -1 by

$$f(x) = \frac{x^3+1}{x+1},$$

then f has a removable discontinuity at -1. Remove it.

2. Sketch the Cartesian graph of the function f given by

$$f(x) = x \cos\left(\frac{1}{x}\right) \quad \text{if } x \neq 0.$$

Is zero a removable discontinuity?

3. Give an example of a bounded function with two non-removable discontinuities.

4. Discuss the discontinuities of the function

$$f(x) = \frac{x^3-8}{x^2-4}.$$

5. Give an example of a function which is continuous on the left at 3 but not continuous at 3.

6. Give an example of a function which is continuous everywhere except at 2 but which is not continuous either on the left or on the right at 2.

7. Give an example of a function, defined over the whole set of of real numbers, which is continuous in the closed interval $[-1, 1]$ but nowhere else.

8. In each of the following cases say whether f is continuous in the closed interval $[0, 1]$.

(i) $f(x) = x^2$.

(ii) $f(x) = [x]$.

(iii) $\begin{cases} f(x) = 1 & \text{if } x \geqslant 0, \\ f(x) = 0 & \text{if } x < 0. \end{cases}$

(iv) $\begin{cases} f(x) = x^2 & \text{if } x > 1, \\ f(x) = 0 & \text{if } 0 < x \leqslant 1, \\ f(x) = -1 & \text{if } x \leqslant 0. \end{cases}$

(v) $\begin{cases} f(x) = \sin \dfrac{1}{x} & \text{if } x \neq 0, \\ f(x) = 0 & \text{if } x = 0. \end{cases}$

9. If f is an increasing function in the open interval (a, b), show that $\lim_{x \to b-} f(x)$ and $\lim_{x \to a+} f(x)$ both exist.

10. Describe the jump discontinuities of the functions defined in question 8.

5.7. Properties of continuous functions

We have now defined continuity and have seen that most commonly occurring functions are continuous. However, the property of continuity will only be of use to us if we are able to prove that some of those properties of a function which we would intuitively expect it to possess from the fact that it is continuous are valid.

The intermediate value theorem. Consider the function $f(x) = x^4 + 3x^3 - 2x^2 - 1$. f is continuous; also $f(0) = -1$, $f(1) = 1$, and $f(-1) = -5$. So part of the Cartesian graph looks more or less as in Fig. 5.12. It is intuitively clear from the graph that the equation

$$x^4 + 3x^3 - 2x^2 - 1 = 0$$

has a solution between zero and 1, i.e. that there is a number u satisfying $0 < u < 1$ and such that $f(u) = 0$. We shall be able formally to draw this conclusion after we have proved the following theorem.

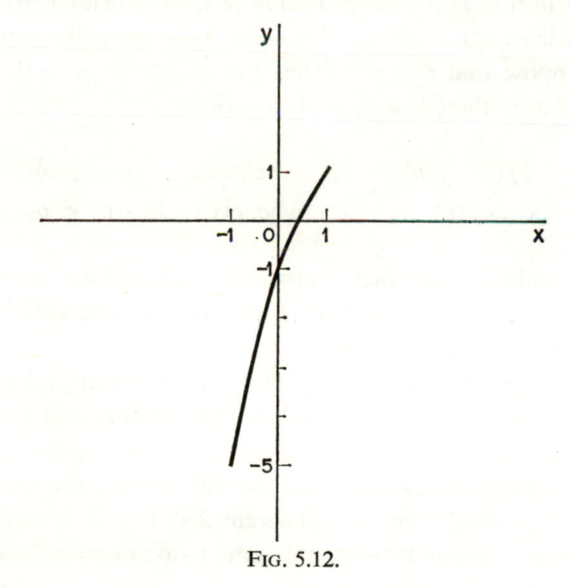

FIG. 5.12.

THEOREM 5.7.1. (The intermediate value theorem). (i) *If f is a function continuous in the closed interval $[a, b]$ and if $f(a) < c$ and $f(b) > c$, then there exists a real number u satisfying $a < u < b$ and such that $f(u) = c$.*

(ii) *If f is a function continuous in the closed interval $[a, b]$ and if $f(a) > c$ and $f(b) < c$, then there exists a real number u satisfying $a < u < b$ and such that $f(u) = c$.*

Proof. (i) Consider the set A of all points x in $[a, b]$ such that $f(x) < c$; i.e.

$$A = \{x : x \in [a, b] \text{ and } f(x) < c\}.$$

A is non-empty since $a \in A$; it is also bounded above by b. So, by Theorem 2.4.2, there is a least upper bound which we shall call u. We shall show that $f(u) = c$.

Suppose first that $f(u) < c$. Then put $\varepsilon = c - f(u) > 0$. Since f is continuous at u, there is a $\delta > 0$ such that

$$|f(x) - f(u)| < \varepsilon \quad \text{whenever} \quad |x - u| < \delta.$$

Thus $f(u + \delta/2) < f(u) + \varepsilon = c$. Therefore $u + \delta/2$ belongs to A and so u is not the least upper bound of A, contrary to our definition of u.

Now suppose that $f(u) > c$. Then put $\varepsilon_1 = f(u) - c > 0$. Since f is continuous at u, there is a $\delta_1 > 0$ such that

$$|f(x) - f(u)| < \varepsilon_1 \quad \text{whenever} \quad |x - u| < \delta_1.$$

Thus $$f(x) > f(u) - \varepsilon_1 = c \quad \text{whenever} \quad |x - u| < \delta_1.$$

Thus no member of the open interval $(u - \delta_1, u)$ belongs to A and so u is not the *least* upper bound of A, contrary to our definition of u. So we conclude that $f(u) = c$.

(ii) Define the function g by $g(x) = -f(x)$. Then $g(a) < -c$ and $g(b) > -c$. Thus by (i) there is a u in $[a, b]$ such that $g(u) = -c$. Then $f(u) = c$. □

We note that in the proof of this theorem we use the completeness property of the real numbers (Theorem 2.4.2). This is essential; the corresponding theorem for rationals, which do not have the completeness property, is not true. We may show this by considering, for example, $f(x) = x^2-2$; $f(1) = -1$, and $f(2) = 2$, but there is no rational u for which $f(u) = 0$.

Continuity and boundedness in a closed interval. Suppose we have a function f which is continuous in the *closed* interval $[a, b]$ (Fig. 5.13). Then $f(a)$ and $f(b)$ have definite finite values, and it is helpful to picture

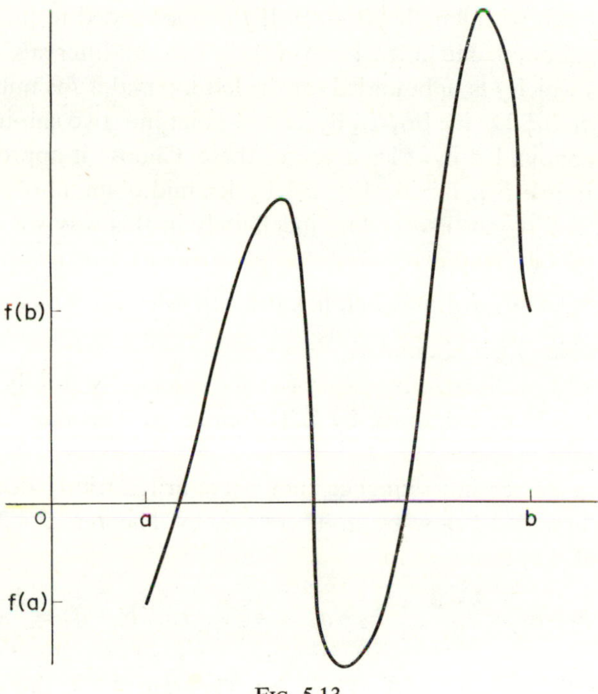

FIG. 5.13.

the function as a continuous piece of string running between the two fixed points $(a, f(a))$ and $(b, f(b))$. Now however many times the function oscillates between a and b and however far up or down it wanders, it is intuitively obvious that the function f is bounded in $[a, b]$ since it is continuous and since its end points are fixed. This we shall now prove.

THEOREM 5.7.2. *If f is a function which is continuous in the* closed *interval $[a, b]$, then f is bounded in $[a, b]$, i.e. there is a positive number M such that $|f(x)| \leqslant M$ for all x in $[a, b]$.*

Proof. We suppose that f is unbounded in $[a, b]$ and prove that this gives a contradiction.

Divide the interval $[a, b]$ by its mid point into two closed sub-intervals (each with length $\frac{1}{2}(b-a)$). If f is unbounded in $[a, b]$, then it will be unbounded in at least one of these two sub-intervals. Call the interval in which f is unbounded (or the left interval if f is unbounded in both) $[a_1, b_1]$. Divide $[a_1, b_1]$ by its mid-point into two sub-intervals. f will be unbounded in at least one of these. Choose it appropriately and call it $[a_2, b_2]$. Divide $[a_2, b_2]$ by its mid-point into two sub-intervals, and so on. Proceeding indefinitely in this way we obtain a sequence of sub-intervals

$$[a_1, b_1], [a_2, b_2], [a_3, b_3], [a_4, b_4], \ldots,$$

in all of which f is unbounded.

Now $\{a_n\}$ is an increasing sequence (not necessarily strictly increasing) and it is bounded above by b. Therefore, by Theorem 3.4.1, $\{a_n\}$ is convergent. Let $\lim a_n = l$.

$\{b_n\}$ is a decreasing sequence (not necessarily strictly decreasing) and it is bounded below by a. Therefore, by Theorem 3.4.2, $\{b_n\}$ is convergent. Let $\lim b_n = m$.

$$\text{Now} \quad b_1 - a_1 = \frac{b-a}{2}, \quad b_2 - a_2 = \frac{b-a}{4}, \ldots, b_n - a_n = \frac{b-a}{2^n}.$$

Therefore $\lim (b_n - a_n) = 0$. Thus, by Theorem 3.3.3, $\lim a_n = \lim b_n = l$. Now $l \geqslant a_n$ for all n and $l \leqslant b_n$ for all n. So l lies in every one of the sub-intervals $[a_n, b_n]$.

Now f is continuous at l so that there is a real number $\delta > 0$ such that

$$|f(x) - f(l)| < 1 \quad \text{whenever} \quad |x - l| < \delta.$$

Therefore f is bounded in the interval $(l-\delta, l+\delta)$. But, since $\lim (b_n - a_n) = 0$, there must be an n such that the interval $[a_n, b_n]$ is completely contained in the interval $(l-\delta, l+\delta)$. Then f is bounded in the interval $[a_n, b_n]$. This gives a contradiction. Thus f is bounded in the closed interval $[a, b]$. □

We note that this result ceases to hold if we relax any of the conditions on the function f.

For example, if f is a function continuous in the open interval (a, b), then f certainly need not be bounded. For consider $f(x) = 1/x$. This is continuous in the open interval $(0, 1)$ but it is not bounded there.

We return to consider once more a function f which is continuous in the closed interval $[a, b]$. Theorem 5.7.2 tells us that f is bounded. Therefore the range of f has a least upper bound and a greatest lower bound. What we now show is that these bounds are actually members of the range of f, or, in other words, that $f(x)$ has a greatest and least value in $[a, b]$. The reader is invited to refer once more to Fig. 5.13 and to convince himself that this result is also intuitively obvious.)

THEOREM 5.7.3. *If f is a function which is continuous in the closed interval $[a, b]$, then it attains its bounds in $[a, b]$; i.e. there is a point u in $[a, b]$ such that $f(u) \geq f(x)$ for all x in $[a, b]$ and there is similarly a point v in $[a, b]$ such that $f(v) \leq f(x)$ for all x in $[a, b]$.*

Proof. Let A be the range of f when the domain of f is $[a, b]$, i.e. A is the set defined by $A = \{f(x) : x \in [a, b]\}$. Then, by Theorem 5.7.2, A is a bounded set. Therefore, by Theorem 2.4.2, there is a least upper bound which we shall call M. Suppose that there is no point u in $[a, b]$ for which $f(u) = M$. Then

$$M - f(x) > 0 \quad \text{for all } x \text{ in } [a, b].$$

Define the function g by $g(x) = 1/\{M - f(x)\}$ for all x in $[a, b]$. By Theorem 5.4.5 g is continuous. Therefore, by Theorem 5.7.2, g is bounded above; in other words, there is a real number K such that $g(x) \leq K$ for all x in $[a, b]$. Thus

$$\frac{1}{M - f(x)} \leq K \quad \text{and so} \quad M - f(x) \geq \frac{1}{K}.$$

Therefore $f(x) \leq M-(1/K)$ for all x in $[a, b]$. Thus $M-(1/K)$ is an upper bound for A and M is not the least upper bound. Since this contradicts our definition of M, we deduce that there is a point u in $[a, b]$ such that $f(u) = M$.

We deduce the second part of the theorem, that the lower bound is attained, from the first part. Define the function h by $h(x) = -f(x)$ for all x in $[a, b]$. Then h is continuous in $[a, b]$ and so, by the first part of this theorem, h attains its upper bound at some point which we shall call v. Then f attains its lower bound at v. □

We note once again that the result of the theorem ceases to hold if we relax any of the conditions on the function f. For example, if f is a continuous function in the open interval (a, b), then, even when f is bounded it need not attain its bounds. The function $f(x) = x$ is bounded and continuous in the open interval $(0, 1)$, but it does not attain either of its bounds within the interval.

Exercises 5.7

1. f and g are functions, both of which are continuous in $[a, b]$. If $f(a)<g(a)$ and if $f(b) > g(b)$, show that there is a point u in $[a, b]$ such that $f(u) = g(u)$. Illustrate the result by means of a Cartesian graph. (We note that this result enables us to conclude with confidence that the equation $x^2 = \cos x$, discussed in the introductory chapter, does have two real roots; see Fig. I.4.)

2. Show that Theorem 5.7.1 is not necessarily true if f is not continuous.

3. Show that the function f defined by

$$f(x) = x^3+x^2$$

is monotonic increasing in the interval $[0, \infty)$. Deduce from the intermediate value theorem that the equation

$$x^3+x^2 = 1$$

has exactly one positive root.

Has the equation a negative root?

4. Give an example of a function which is continuous in the *open* interval $(0, 1)$, is bounded in this interval and attains its bounds within this interval.

5. Give an example of a function which is discontinuous in the *closed* interval $[0, 1]$, is bounded in this interval but which does not attain either of its bounds.

6. Find the points at which the continuous function f attains its bounds in the closed interval [0, 4] in each of the following cases:

(i) $f(x) = x^3$. (ii) $f(x) = 3-x^2$. (iii) $f(x) = \sin x$.

7. Give an example of a function f for which $f(a) < f(b)$, which is *not* continuous in $[a, b]$ but which is such that, if any c is chosen satisfying $f(a) < c < f(b)$, then a point u may be found with $f(u) = c$.

Exercises for Chapter 5

1. State carefully what you mean by the statements (a) $\lim_{x \to a} f(x) = L$, (b) the function f is continuous at $x = a$.

Prove that if the function f is continuous at $x = a$, and the function g is continuous at $x = f(a)$, then the composite function gf is continuous at $x = a$.

Let p and q be the functions given by:

$$p(x) = [x], \text{ the integral part of } x, \text{ for all real } x.$$

$$q(x) = \begin{cases} 0, & \text{when } x \text{ is an integer,} \\ x^2, & \text{when } x \text{ is real but not integral.} \end{cases}$$

Show that:

(a) $\lim_{x \to 1} q(x)$ exists, but q is not continuous at $x = 1$;

(b) $\lim_{x \to 1} p(x)$ does not exist and p is not continuous at $x = 1$;

(c) the composite function qp is continuous for all real x. [T.C]

2. Give strict definitions of the limit of a sequence and the continuity of a function at a point.

If $\lim_{n \to \infty} a_n = a$ and $f(x)$ is continuous at the point a, prove that $\lim_{n \to \infty} f(a_n) = f(a)$.

If $\lim_{n \to \infty} f(a_n) = f(a)$ for *all* sequences a_n such that $\lim_{n \to \infty} a_n = a$, prove that $f(x)$ is continuous at a. [T.C.]

3. (i) State the values of the following limits where they exist:

$$\text{(a)} \lim_{x \to 0} \sin\left(\frac{1}{x}\right), \quad \text{(b)} \lim_{x \to 0} \left(\frac{\sin x}{x}\right), \quad \text{(c)} \lim_{x \to 0} \frac{\log(1+x)}{x}.$$

(ii) The sequence $\{a_n\}$ is given by $a_n = n/(n+1)+\cos(n\pi/3)$, where $n = 1, 2, 3\ldots$.

Determine:

(a) the largest number m and the smallest number M such that $m \leqslant a_n \leqslant M$, for, all n,

(b) $\lim_{n\to\infty} a_{6n}$, $\lim_{n\to\infty} a_{6n+1}$

(iii) Given that

$$f(x) = \lim \frac{1}{1+(x/3)^{2n}},$$

sketch the graph of $f(x)$ for $|x| < 4$. [B.Ed.]

4. What is meant by saying that a function is monotonic?

By applying the intermediate value theorem to $f(x) = x^3+3x^2$, show that the equation $x^3+3x^2 = 2$ has one and only one real root in [0, 1].

The continuous function g maps [0,1] one-to-one on to [0, 1], $g(0) = 0$ and $g(1) = 1$. Prove that g is strictly monotonic increasing. [B.Ed.]

Other questions involving the subject-matter contained in this chapter and the following chapter will be found in the exercises for Chapter 6.

CHAPTER 6

THE DERIVATIVE

6.1. Derivatives and their evaluation

THE previous chapter concerned functions of a real variable. We discussed, among other things, increasing functions and continuous functions. Both these properties of functions tell us something about how $f(x)$ changes when x changes, but neither tell us the rate at which $f(x)$ is changing.

Consider the function $f(x) = 2x$. f is an increasing function, so that we shall call its rate of change positive. The function f is also continuous. If we draw a parallel axis graph of f, we see that any interval in the domain is mapped by f into an interval of twice its length in the codomain (Fig. 6.1). Thus the function f produces an enlargement with scale factor 2.

Alternatively we may consider the rate of change of the function f in the following way. If the point P moves continuously along the domain axis and if Q is the image of P under the function f, then Q moves twice as fast along the codomain axis.

If we now draw the Cartesian graph of the function f, we see that the gradient of the graph is constant and is equal to 2, i.e. $\tan \theta = 2$ (Fig. 6.2).

Whichever point of view we adopt, we see that the rate of change of the function f is constant and is equal to 2; this is because

$$\frac{f(b)-f(a)}{b-a} = 2$$

whatever values of a and b we may choose, provided $a \neq b$.

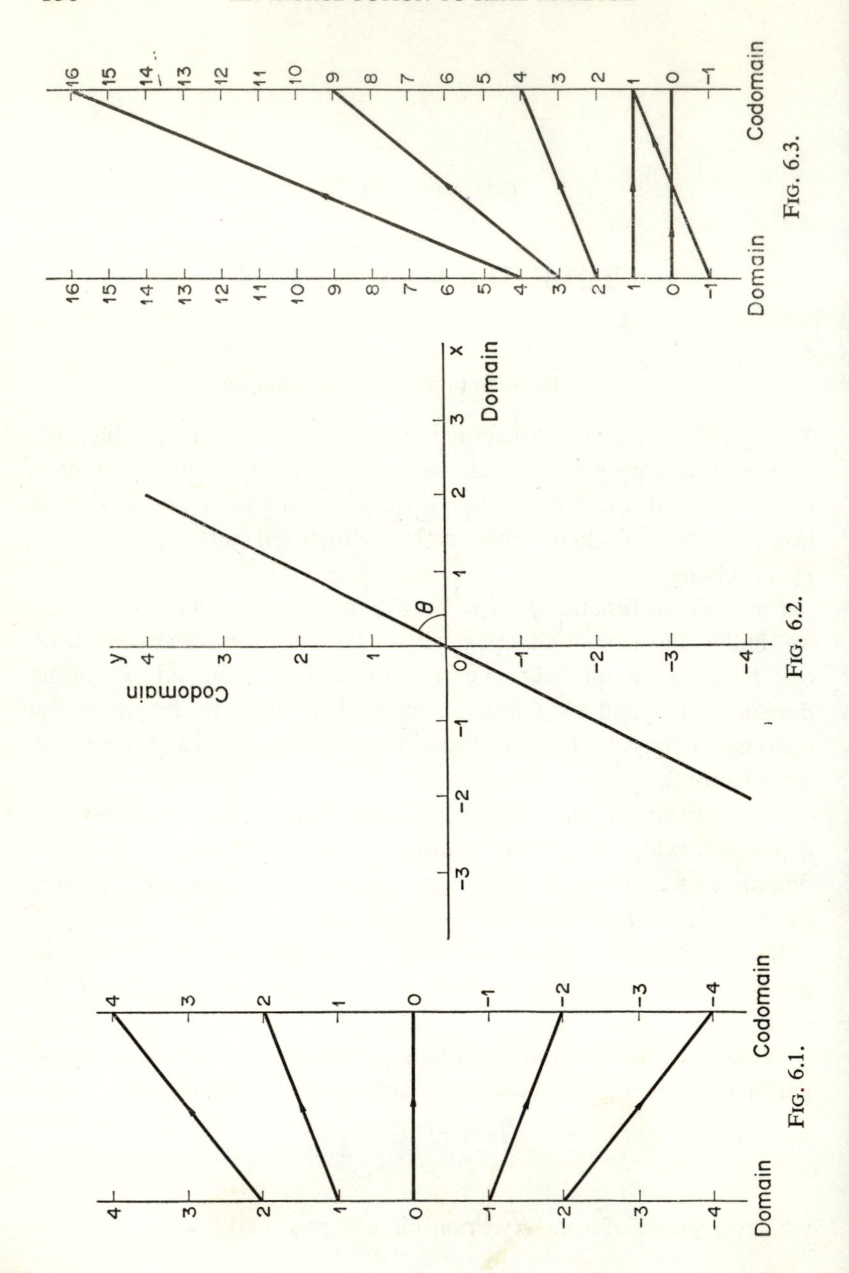

FIG. 6.1.

FIG. 6.2.

FIG. 6.3.

Consider now the function g defined by $g(x) = x^2$. This is a function which is increasing in $[0, \infty)$ and decreasing in $(-\infty, 0]$. Thus we shall call its rate of change positive in $[0, \infty)$ and negative in $(-\infty, 0]$. g is also continuous. If we examine the parallel axis graph of g (Fig. 6.3) we see that the magnification produced by g is not constant. Some intervals are enlarged more than others. For example, the interval $[0, 1]$ maps on to the interval $[0, 1]$ which is an enlargement with scale factor 1, whereas the interval $[3, 4]$ maps on to the interval $[9, 16]$ which is an enlargement with scale factor 7. Thus we may speak of the (average) enlargement of a particular interval, although we are aware that the enlargement is not constant throughout the interval.

Alternatively, if the point P moves continuously along the domain axis and if Q is the image of P under the function g, then Q sometimes moves slower than P, sometimes faster, and sometimes in the opposite direction. If P moves between two given points, then we may measure the average ratio of the speed of Q to that of P; this will, however, differ from the ratio at an instant.

If we now draw the Cartesian graph of the function (Fig. 6.4), we

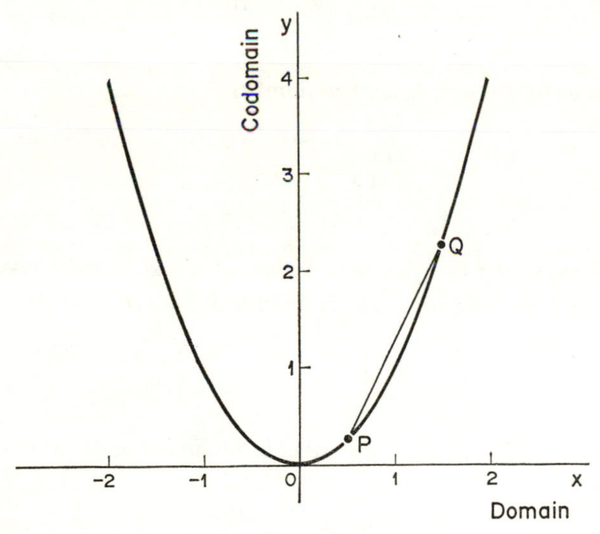

FIG. 6.4.

see that the gradient of this graph is not constant. We can readily measure the average gradient of the graph between the points A and B, this being the gradient of the chord AB. This, however, is not the gradient of the graph at a particular point.

Whichever point of view we adopt, we see that the average rate of change of the function g between the points $x = a$ and $x = b$ is given by

$$\frac{g(b)-g(a)}{b-a},$$

but this ratio is not independent of the choice of a and b.

The average rate of change of the function g between the points $x = a$ and $x = b$ may or may not give a good representation of the actual rate of change of g at $x = a$. Whether it does or not will depend on how much the rate of change varies in $[a, b]$. Nevertheless, the nearer b approaches a the more likely it is that the rate of change of g at $x = a$ will be well represented by its average rate of change in $[a, b]$.

This leads us to represent the rate of change of the function g at the point $x = a$ by the number

$$\lim_{b \to a} \left[\frac{g(b)-g(a)}{b-a}\right]$$

or, if we write $b-a = h$, by the number

$$\lim_{h \to 0} \left[\frac{g(a+h)-g(a)}{h}\right].$$

DEFINITION. *If f is a function defined at a and in some open interval containing a, then we say that f is* differentiable *at a provided*

$$\lim_{h \to 0} \left[\frac{f(a+h)-f(a)}{h}\right]$$

exists and is finite. In this case we call the limit the derivative *at a and we write*

$$f'(a) = \lim_{h \to 0} \left[\frac{f(a+h)-f(a)}{h}\right].$$

Generally f is differentiable at most or all points of the domain, so that f' is a function whose domain is the domain of f or some subset.

DEFINITION. *If f' is the function defined by*

$$f'(x) = \lim_{h \to 0} \left[\frac{f(x+h)-f(x)}{h} \right],$$

then f' is called the derived function *or the* derivative *of f.*

It is intuitively obvious that f will not have a well-defined, finite rate of change at the point a unless f is changing continuously at the point a.

THEOREM 6.1.1. *If f is differentiable at a then it is continuous at a.*

Proof. If f is differentiable at a, then

$$\lim_{h \to 0} \left[\frac{f(a+h)-f(a)}{h} \right]$$

exists. Therefore we can find a real number $\delta_0 > 0$ such that

$$\left| \frac{f(a+h)-f(a)}{h} - f'(a) \right| < 1 \quad \text{whenever} \quad 0 < |h| < \delta_0.$$

Thus $|\{f(a+h)-f(a)\}/h| < 1+|f'(a)|$ whenever $0 < |h| < \delta_0$. Given $\varepsilon > 0$, put $\delta = \min(\delta_0, \varepsilon/|1+f'(a)|.)$. Then

$$|f(a+h)-f(a)| < |h|(1+|f'(a)| < \varepsilon \quad \text{whenever} \quad |h| < \delta.$$

Therefore f is continuous at a. □

We note that the converse of this theorem is not true. Functions do exist which are continuous at a point without being differentiable there. For example, if $f(x) = |x|$, then it may readily seen from a sketch graph or proved formally that f is continuous at zero but not differentiable there. More extreme examples do exist; functions have been constructed which are everywhere continuous and nowhere differentiable.

The sum, product, and quotient formulae. The following results give us practical methods of finding the derivatives of many commonly occurring functions.

THEOREM 6.1.2. *f and g are functions both differentiable at a.*

(i) *If F is defined by $F(x) = f(x) \pm g(x)$ then f is differentiable and $F'(a) = f'(a) \pm g'(a)$.*

(ii) *If G is defined by $G(x) = F(x)\, G(x)$ then G is differentiable at a and $G'(a) = f(a)\, g'(a) + f'(a)\, g(a)$.*

(iii) *If H is defined by $H(x) = f(x)/g(x)$ then H is differentiable at a and*

$$H'(a) = \frac{g(a)f'(a) - f(a)g'(a)}{[g(a)]^2}.$$

Proof. (i)

$$\begin{aligned} F'(a) &= \lim_{h \to 0} \left[\frac{F(a+h) - F(a)}{h} \right] \\ &= \lim_{h \to 0} \left[\frac{f(a+h) - f(a)}{h} \pm \frac{g(a+h) - g(a)}{h} \right] \\ &= \lim_{h \to 0} \left[\frac{f(a+h) - f(a)}{h} \right] \pm \lim_{h \to 0} \left[\frac{g(a+h) - g(a)}{h} \right] \\ &= f'(a) \pm g'(a). \end{aligned}$$

(ii)

$$\begin{aligned} G'(a) &= \lim_{h \to 0} \left[\frac{G(a+h) - G(a)}{h} \right] \\ &= \lim_{h \to 0} \left[\frac{g(a+h)[f(a+h) - f(a)] + f(a)[g(a+h) - g(a)]}{h} \right] \\ &= \lim_{h \to 0} [g(a+h)] \lim_{h \to 0} \left[\frac{f(a+h) - f(a)}{h} \right] \\ &\quad + f(a) \lim_{h \to 0} \left[\frac{g(a+h) - g(a)}{h} \right]. \end{aligned}$$

By Theorem 6.1.1, g is continuous at a, so that

$$\lim_{h\to 0} g(a+h) = g(a).$$

Therefore $$G'(a) = g(a)f'(a)+f(a)\,g'(a).$$

$$\begin{aligned}
\text{(iii) } H'(a) &= \lim_{h\to 0}\left[\frac{H(a+h)-H(a)}{h}\right] \\
&= \lim_{h\to 0}\left[\frac{f(a+h)\,g(a)-g(a+h)\,f(a)}{hg(a)\,g(a+h)}\right] \\
&= \lim_{h\to 0}\left[\frac{[f(a+g)-f(a)]\,g(a)-[g(a+h)-g(a)]\,f(a)}{hg(a)\;g(a+h)}\right] \\
&= \lim_{h\to 0}\left[\frac{f(a+h)-f(a)}{h}\right]\lim_{h\to 0}\left[\frac{g(a)}{g(a)\,g(a+h)}\right] \\
&\quad -\lim_{h\to 0}\left[\frac{g(a+h)-g(a)}{h}\right]\lim_{h\to 0}\left[\frac{f(a)}{g(a)\,g(a+h)}\right] \\
&= \frac{f'(a)\,g(a)-g'(a)f(a)}{[g(a)]^2}. \;\square
\end{aligned}$$

Composite functions (function of a function). In the previous chapter we saw that, if we want to prove that f is continuous when f is a function like

$$f(x) = \sqrt{(x^2+x+1)},$$

then it is helpful to consider f as built from the functions F and G, where

$$F(x) = x^2+x+1 \quad \text{and} \quad G(x) = \sqrt{x}.$$

Such a breakdown is also useful in obtaining the derivative of the function f.

More generally, if f is any function built from the differentiable functions F and G, so that $f = G \circ F$, then the local magnification produced by f at the point a is seen intuitively to equal the product of the local magnification produced by F at the point a and the local magnification produced by G at the point $F(a)$ (Fig. 6.5). We may express this result in symbols:

$$f'(a) = F'(a)\,G'\big(F(a)\big).$$

The formal proof of this result, which is sometimes called the chain rule, is essentially as simple as the intuitive idea, but there is a technical

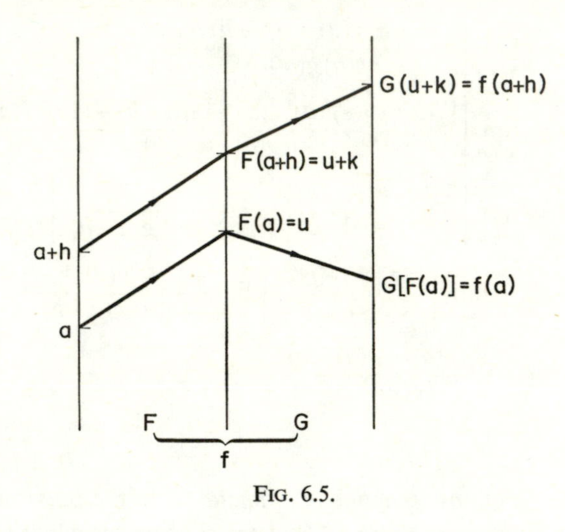

FIG. 6.5.

complication. We smooth the way to some extent by proving first a lemma.

LEMMA. *If* $f(x) \to l$ *as* $x \to a$ *and if* $l \neq 0$, *then we can find a real number* $\delta > 0$, *such that* $f(x) \neq 0$ *for all* x, *not equal to* a, *in the open interval* $(a-\delta, a+\delta)$.

Proof. Since $f(x) \to l$ we may find a real number $\delta > 0$ such that

$$|f(x)-1| < \left|\frac{l}{2}\right| \quad \text{whenever} \quad 0 < |x-a| < \delta.$$

Therefore $f(x) \neq 0$ for all x, not equal to a, in $(a-\delta, a+\delta)$. □

THEOREM 6.1.3. (the chain rule). *If* $f = G \circ F$ *and if* $F(a) = u$ *and if* F *is differentiable at* a *and* G *is differentiable at* u, *then* f *is differentiable at* a *and*

$$f'(a) = G'(u)\, F'(a).$$

Proof. Suppose first that $F'(a) \neq 0$. Then, by the lemma, we may find a real number $\delta > 0$, such that

$$F(a+h)-F(a) \neq 0 \quad \text{if} \quad 0 < |h| < \delta.$$

Put $u+k = F(a+h)$. Then we have $f(a) = G(u)$ and $f(a+h) = G(u+k)$. If $0 < |h| < \delta$,

$$\begin{aligned} f'(a) &= \lim_{h\to 0}\left[\frac{f(a+h)-f(a)}{h}\right] \\ &= \lim_{h\to 0}\left[\frac{G(u+k)-G(u)}{h}\right] \\ &= \lim_{h\to 0}\left[\frac{G(u+k)-G(u)}{k}\right] \lim\left[\frac{k}{h}\right]. \end{aligned}$$

Now $k \to 0$ as $h \to 0$ since F is continuous at a (by Theorem 6.1.1). Thus $f'(a) = G'(u)\, F'(a)$.

Now suppose that $F'(a) = 0$. Then, for some values of h, k may be zero, giving

$$\frac{f(a+h)-f(a)}{h} = \frac{G(u+k)-G(u)}{h} = \frac{G(u)-G(u)}{h} = 0.$$

For other values of h, k may be non-zero, giving

$$\frac{f(a+h)-f(a)}{h} = \frac{G(u+k)-G(u)}{k}\,\frac{k}{h}.$$

Now $\lim k/h = F'(a) = 0$. Therefore, whether or not k is zero,

$$f'(a) = \lim_{h\to 0}\left[\frac{f(a+h)-f(a)}{h}\right] = 0.$$

So, in this case too, $f'(a) = G'(u)\, F'(a)$. □

Inverse functions. Consider the function f which is one to one and which therefore has an inverse function, say F. Suppose that $f(a) = u$. If f is represented on a parallel axis graph, then F may be represented simply by reversing the directions of the arrows of the graph of f, so that the domain becomes the codomain and vice versa. It is then clear that the local magnification of F at u is the reciprocal of the local magnification of f at a. Thus we have the following theorem.

THEOREM 6.1.4. *f is a one to one function with inverse F. If f is differentiable at a and if $f'(a) \neq 0$, then F is differentiable at u and*

$$F'(u) = \frac{1}{f'(a)}.$$

Proof. Put $g = F \circ f$ so that $g(x) = x$. Then $g'(a) = 1$. Thus, by Theorem 6.1.3, $F'(u)f'(a) = 1$. Therefore

$$F'(u) = \frac{1}{f'(a)}. \quad \square$$

The derivatives of some commonly occurring functions

THEOREM 6.1.5. *If n is a natural number and if $f(x) = x^n$, then f is differentiable everywhere and*

$$f'(a) = na^{n-1}.$$

Proof. $$\frac{(a+h)^n - a^n}{h} = \frac{(a+h-a)\left((a+h)^{n-1} + (a+h)^{n-2}a + \ldots + a^{n-1}\right)}{h}$$
$$= (a+h)^{n-1} + (a+h)^{n-2}a + \ldots + a^{n-1}.$$

Therefore $$f'(a) = \lim_{h \to 0} [(a+h)^{n-1} + (a+h)^{n-2}a + \ldots + a^{n-1}]$$
$$= na^{n-1}. \quad \square$$

THEOREM 6.1.6. *If n is a natural number, if a is a real number greater than zero and if*

$$f(x) = x^{1/n},$$

then f is differentiable at a and

$$f'(a) = \frac{1}{n} a^{1/n-1}.$$

Proof. f is one to one. Let F be its inverse function. Then $F(x)=x^n$. Also $f(a) = a^{1/n}$. Therefore, by Theorem 6.1.4,

$$f'(a) = \frac{1}{F'(a^{1/n})} = \frac{1}{n(a^{1/n})^{n-1}} = \frac{1}{na^{1-1/n}}$$
$$= \frac{1}{n} a^{1/n-1}.$$

THEOREM 6.1.7. *If p and q are natural numbers, if a is a real number greater than zero, and if*

$$f(x) = x^{p/q},$$

then f is differentiable at a and

$$f'(a) = \frac{p}{q} x^{p/q-1}.$$

Proof. Put $f = G \circ F$ where

$$F(x) = x^p \quad \text{and} \quad G(x) = x^{1/q}.$$

Then, by Theorem 6.1.3, f is differentiable at a and

$$f'(a) = G'(a^p)\, F'(a) = \frac{1}{q} (a^p)^{1/q-1} \cdot pa^{p-1} = \frac{p}{q} a^{p/q-p+p-1}$$
$$= \frac{p}{q} a^{p/q-1}. \quad \square$$

THEOREM 6.1.8. *If p and q are natural numbers, if a is a real number greater than zero, and if*

$$f(x) = \frac{1}{x^{p/q}},$$

then f is differentiable at a and

$$f'(a) = -\frac{p}{q}\,\frac{1}{a^{p/q+1}}\,.$$

Proof. By Theorem 6.1.2(iii),

$$f'(a) = \frac{(-p/q)a^{p/q-1}}{a^{2p/q}} = -\frac{p}{q}\,\frac{1}{a^{p/q+1}}\,.$$

The results of Theorems 6.1.5–6.1.8 may be summarized by saying that if r is any rational and if a is a real number greater than zero and if $f(x) = x^r$ then f is differentiable at a and $f'(a) = ra^{r-1}$.

THEOREM 6.1.9. (i) *If $f(x) = \sin x$, then f is differentiable everywhere and $f'(a) = \cos a$.*

(ii) *If $f(x) = \cos x$ then f is differentiable everywhere and $f'(a) = -\sin a$.*

Proof. (i) $f'(a) = \lim\limits_{h\to 0}\left[\dfrac{f(a+h)-f(a)}{h}\right]$

$$= \lim_{h\to 0}\left[\frac{2\cos(a+h/2)\sin h/2}{h}\right]$$

$$= \lim_{h\to 0}\left[\cos\left(a+\frac{h}{2}\right)\right]\lim_{h\to 0}\left[\frac{\sin h/2}{h/2}\right]$$

$$= \cos a\cdot 1 = \cos a.$$

(ii) $f'(a) = \lim\limits_{h\to 0}\left[\dfrac{f(a+h)-f(a)}{h}\right]$

$$= \lim_{h\to 0}\left[\frac{-2\sin(a+h/2)\sin h/2}{h}\right]$$

$$= \lim_{h\to 0}\left[-\sin\left(a+\frac{h}{2}\right)\right]\lim_{h\to 0}\left[\frac{\sin h/2}{h/2}\right]$$

$$= -\sin a\cdot 1 = -\sin a. \ \square$$

The reader will probably be familiar with the use of Theorems 6.1.2–6.1.4 to obtain the derivatives of other functions. For example:

(i) If $f(x) = x^2 \sin x$, then we use Theorem 6.1.2(ii) to obtain

$$f'(x) = x^2 \cos x + 2x \sin x.$$

(ii) if $f(x) = \tan x = \sin x/\cos x$, then, when $x \neq (n+\frac{1}{2})\pi$, we use Theorem 6.1.2(iii) to obtain

$$f'(x) = \frac{\cos^2 x + \sin^2 x}{\cos^2 x} = \sec^2 x.$$

(iii) if $f(x) = \sin^3 x = (\sin x)^3$, then we use Theorem 6.1.3 to obtain

$$f'(x) = 3 \sin^2 x \cos x.$$

Exercises 6.1

1. If $f(x) = c$, where c is a constant, show that f is differentiable everywhere and that $f'(a) = 0$ for all a.

2. If $f(x) = |x|$ show that f is not differentiable at zero.

3. If $f(x) = x^3$ show from first principles that f is differentiable at the point 2 and find its derivative there.

4. The function f is defined by

$$\begin{cases} f(x) = x^2 & \text{if} \quad x \geqslant 0, \\ f(x) = 0 & \text{if} \quad x < 0. \end{cases}$$

Show that f is differentiable at the point zero.

5. The function f is defined by

$$\begin{cases} f(x) = x^2 & \text{if} \quad x \geqslant 1, \\ f(x) = 2x-1 & \text{if} \quad 0 \leqslant x < 1, \\ f(x) = x-1 & \text{if} \quad -1 \leqslant x < 0, \\ f(x) = x^2 & \text{if} \quad x < -1. \end{cases}$$

Sketch the Cartesian graph of f. At which points is f (i) not continuous, (ii) not differentiable?

6. Use Theorems 6.1.2 and 6.1.3 to find the derivative of f in each of the following cases:

(i) $f(x) = x^3 \cos x$.

(ii) $f(x) = \dfrac{\sin x}{x}$

(iii) $f(x) = \sin\left(\dfrac{1}{x}\right)$ if $x \neq 0$

(iv) $f(x) = \cos^3 x$.

(v) $f(x) = \sin^2(3x)$.

(vi) $f(x) = x^3 \tan^2 x$.

6.2. Rolle's theorem and the nature of the derivative

Suppose we have a function and draw its Cartesian graph. If u is a point such that

$$f(u) \geqslant f(x)$$

for all x and if we are able to draw a unique tangent at u (Fig. 6.6), then it is evident geometrically that the tangent will be horizontal. The same applies at a point v such that

$$f(v) \leqslant f(x).$$

We prove this result in the following theorem.

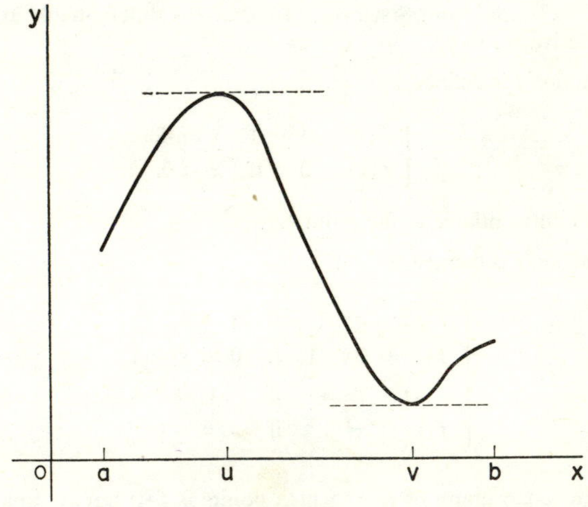

FIG. 6.6.

THEOREM 6.2.1. *f is defined in the open interval (a, b), u is a point of (a, b) and f is differentiable at u. If either*

$$\text{(i)} \quad f(u) \geqslant f(x) \quad \textit{for all } x \textit{ in} \quad (a, b),$$

or

$$\text{(ii)} \quad f(u) \leqslant f(x) \quad \textit{for all } x \textit{ in} \quad (a, b),$$

then $f'(u) = 0$.

Proof. (i) If $h > 0$ and if $u+h$ belongs to (a, b)

$$\frac{f(u+h)-f(u)}{h} < 0.$$

Therefore

$$f'(u) = \lim_{h \to 0} \left[\frac{f(u+h-f(u)}{h} \right] \leqslant 0.$$

If $h < 0$ and if $u+h$ belongs to (a, b),

$$\frac{f(u+h)-f(u)}{h} > 0.$$

Therefore

$$f'(u) = \lim_{h \to 0} \left[\frac{f(u+h)-f(u)}{h} \right] \geqslant 0$$

Therefore $f'(u) = 0$.

The proof of (ii) is similar and is left to the reader. □

Suppose now we have a function f, continuous in the closed interval $fa, b]$ and differentiable in the open interval (a, b). Suppose also that $[(a) = f(b)$. A graph of one such function is shown in Fig. 6.7. Then it is evident geometrically that there is a point between a and b at which the tangent to the graph is horizontal. This result is known as Rolle's theorem.

THEOREM 6.2.2. (Rolle's theorem). *If the function f is continuous in the closed interval $[a, b]$, differentiable in the open interval (a, b) and if $f(a) = f(b)$, then there is a point u in (a, b) such that $f'(u) = 0$.*

Proof. Suppose first that f is constant. Then $f'(x) = 0$ for all x in (a, b) and the theorem is proved.

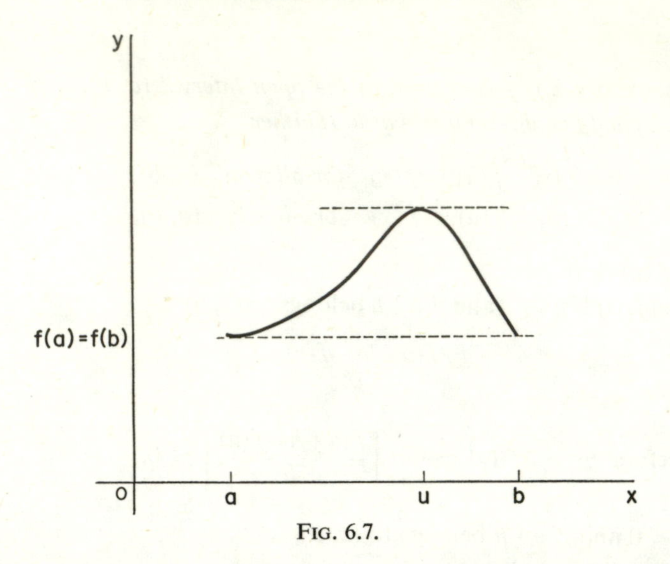

FIG. 6.7.

Suppose now that f is not constant. Since f is continuous in $[a, b]$, then it is bounded by Theorem 5.7.2. Since f is not constant, the upper bound is different from the lower bound. By Theorem 5.7.3, f attains both of these bounds in $[a, b]$. Since $f(a) = f(b)$, f must attain at least one of its bounds in (a, b), say at the point u.

Then, by Theorem 6.2.1, $f'(u) = 0$. □

Rolle's theorem is a crucial theorem of the calculus, enabling many significant results to be proved. Most of the remaining theorems of this chapter depend for their proofs upon Rolle's theorem.

In Theorem 6.1.1 we saw that if the function f is differentiable then it must be continuous. But what of the derivative f'; is f' always continuous?

Consider the function f defined by

$$\begin{cases} f(x) = x^2 \sin 1/x & \text{if} \quad x \neq 0, \\ f(0) = 0. \end{cases}$$

We first show that f is differentiable at zero.

$$f'(0) = \lim_{h \to 0} \left[\frac{f(h) - f((0)}{h} \right] = \lim_{h \to 0} \left[\frac{h^2 \sin 1/h}{h} \right] = 0.$$

Thus the derived function f' is given by

$$\begin{cases} f'(x) = 2x \sin \dfrac{1}{x} - \cos \dfrac{1}{x} & \text{if} \quad x \neq 0, \\ f'(0) = 0. \end{cases}$$

f' is not continuous at zero since (as we have seen) $\cos 1/x$ does not tend to a limit as $x \to 0$. Therefore a derivative is not always continuous.

One of the most important properties of a continuous function is the intermediate value property (Theorem 5.7.1). Even though derivatives are not always continuous, they do always possess the intermediate value property, as the following two theorems show.

THEOREM 6.2.3. (Darboux's theorem). *If f is a function, differentiable in the closed interval $[a, b]$ and if either*

(i) $f'(a) < 0$ *and* $f'(b) > 0$ or (ii) $f'(a) > 0$ *and* $f'(b) < 0$,

then there is a point u in (a, b) such that $f'(u) = 0$.

Proof. (i) $f'(a) < 0$ and $f'(b) > 0$. Suppose, first, that $f(a) < f(b)$ (the reader will find a sketch graph of the situation helpful). Since $f'(a) < 0$,

$$\frac{f(a+h)-f(a)}{h} < 0$$

for all sufficiently small h. Therefore there is a point $a+h$ in (a, b) for which $f(a+h)-f(a) < 0$. Put $v=a+h$. Then $f(v) < f(a)$. Now f is continuous in $[v, b]$ and $f(v) < f(a) < f(b)$. Therefore, by the intermediate value theorem, there is a point w in (v, b) such that $f(w)=f(a)$. Also f is continuous in $[a, w]$ and differentiable in (a, w). Thus, by Rolle's theorem, there is a point u in (a, w) such that $f'(u) = 0$. Suppose now that $f(a) > f(b)$. Since $f'(b) > 0$,

$$\frac{f(b+h)-f(b)}{h} > 0$$

for all sufficiently small h. Therefore there is a point $b+h$ (with h negative) in (a, b) for which $f(b+h)-f(b) < 0$. Put $v = b+h$. Then $f(v) < f(b)$. Now f is continuous in $[a, v]$ and $f(v) < f(b) < f(a)$. Therefore, by the intermediate value theorem, there is a point w in (a, v) such that $f(w) = f(b)$. Also f is continuous in $[w, b]$ and differentiable in (w, b). Thus, by Rolle's theorem, there is a point u in (w, b) such that $f'(u) = 0$.

The proof of (ii) is similar and is left to the reader. □

THEOREM 6.2.4. *If f is a function, differentiable in the closed interval $[a, b]$ and if either*

$$\text{(i) } f'(a) < c < f'(b) \quad \text{or} \quad \text{(ii) } f'(a) > c > f'(b),$$

then there is a point u in (a, b) such that $f'(u) = 0$.

Proof. (i) If we define the function g by $g(x) = f(x)-cx$, then $g'(a) = f'(a)-c < 0$ and $g'(b) = f'(b)-c > 0$. Thus g satisfies the conditions of Theorem 6.2.3(i). Hence there is a point u in (a, b) such that $g'(u) = 0$. But $g'(u) = f'(u)-c$ so that $f'(u) = c$.

The proof of (ii) follows similarly from Theorem 6.2.3(ii). □

In view of Theorem 6.2.4 derivatives are sometimes described as being Darboux continuous, not because they are always continuous, but because they always possess the intermediate value property characteristic of continuous functions.

Exercises 6.2

1. Prove Theorem 6.2.3. (ii)

2. Prove Theorem 6.2.4 (ii).

3. If $f(x) = x^2-4x+1$, show that f satisfies the conditions of Rolle's theorem in the interval $[1, 3]$ and find a point u in $(1, 3)$ such that $f'(u) = 0$.

4. If $f(x) = x^4+x^3-5x-7$, show that there is a point u in the interval $(0, 1)$ such that $f'(u) = 0$.

6.3. Mean value theorems

Figure 6.8 shows the Cartesian graph of a function f. If f is continuous in $[a, b]$ and differentiable in (a, b), then it is evident that we may find a point R between P and Q such that the tangent at R is parallel to the chord PQ. Alternatively, we may express this result by saying

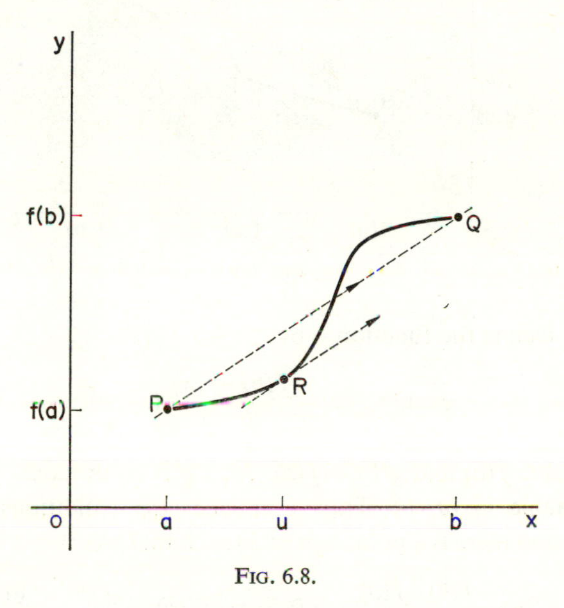

FIG. 6.8.

that there is a point at which the rate of change of f is equal to its average rate of change in $[a, b]$. This result is known as the first mean value theorem.

THEOREM 6.3.1 (the first mean value theorem). *If the function f is continuous in $[a, b]$ and differentiable in (a, b), then there is a point u in (a, b) such that*

$$f'(u) = \frac{f(b)-f(a)}{b-a}.$$

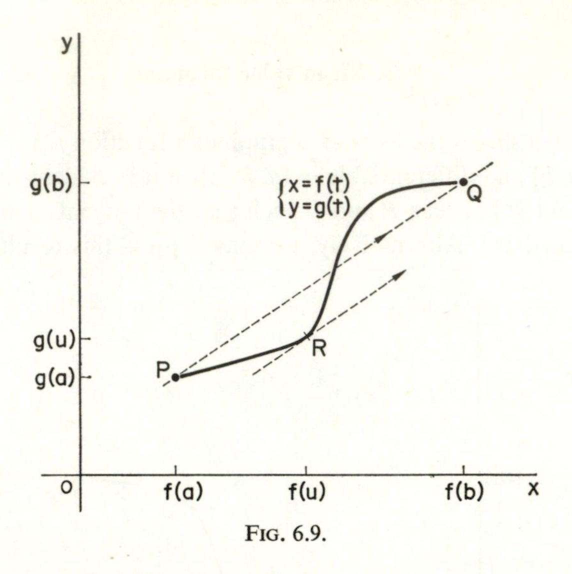

FIG. 6.9.

Proof. Define the function g by

$$g(x) = f(x) - \frac{f(b)-f(a)}{b-a}(x-a).$$

Then $g(a) = f(a)$ and $g(b) = f(a)$. Also g is continuous in $[a,b]$ and differentiable in (a, b). Thus g satisfies the conditions of Rolle's theorem and there is a point u in (a, b) such that $g'(u) = 0$. Thus

$$f'(u) - \frac{f(b)-f(a)}{b-a} = 0 \quad \text{or} \quad f'(u) = \frac{f(b)-f(a)}{b-a}. \quad \square$$

2

Suppose that the curve shown in Fig. 6.9 is described parametrically by the equations

$$x = f(t),$$
$$y = g(t).$$

Then the curve may also be described by the function $y = F(x)$, where $F = gf^{-1}$. Hence we deduce from Theorems 6.1.3 and 6.1.4 that

$$F'(x) = \frac{g'(t)}{f'(t)}.$$

If we now choose R so that the gradient of the tangent at R is parallel to the chord PQ, and if R is the point $(f(u), g(u))$ then the gradient of the tangent at R is $g'(u)/f'(u)$ and the gradient of PQ is $\{g(b)-g(a)\}/\{f(b)-f(a)\}$. This suggests the following theorem.

THEOREM 6.3.2 (Cauchy's mean value theorem). *If the functions f and g are both continuous in $[a, b]$ and both differentiable in (a, b) and if f' is non-zero throughout (a, b), then there is a point u in (a, b) such that*

$$\frac{g'(u)}{f'(u)} = \frac{g(b)-g(a)}{f(b)-f(a)}.$$

Proof. Since f' is non-zero throughout (a, b), Rolle's theorem assures us that $f(a) \neq f(b)$. Define the function G by

$$G(x) = g(x) - \frac{g(b)-g(a)}{f(b)-f(a)}(f(x) - f(a)).$$

Then $G(a) = g(a)$ and $G(b) = g(a)$. Also G is continuous in $[a, b]$ and differentiable in (a, b). Thus G satisfies the conditions of Rolle's theorem and there is a point u in (a, b) such that $g'(u) = 0$. Thus

$$g'(u) - \frac{g(b)-g(a)}{f(b)-f(a)} f'(u) = 0$$

and, since $f'(u) \neq 0$, we may divide by $f'(u)$ to obtain

$$\frac{g'(u)}{f'(u)} = \frac{g(b)-g(a)}{f(b)-f(a)}.$$

Exercises 6.3

1. If $f(x) = x^2$, find the value of u when the first mean value theorem is applied to f in the interval $[-1, 3]$.

2. If $f(x) = x^2$ show that the point u, when the first mean value theorem is applied to f in the interval $[a, b]$, is always the mid-point of $[a, b]$.

3. If $f(x) = 1/x$, find the value of u when first mean value theorem is applied to f in the interval $[-5, -2]$.

4. By considering the function $f(x) = x^4+x^3+3x^2$, use the first mean value theorem to show that the equation

$$4x^3+3x^2+6x = 27$$

has a solution in the interval $(1, 3)$.

5. Use the first mean value theorem to show that the equation

$$5 \cos x+4 \sin 4x= \frac{3}{\pi}$$

has a solution in the interval $(\pi/6, \pi/2)$.

6. Use Cauchy's mean value theorem to show that the equation

$$\frac{\cos x}{x} = \frac{8}{\pi^2}$$

has a solution in the interval $(0, \pi/2)$.

7. If the functions f and g are defined by

$$f(x) = x^2 \quad \text{and} \quad g(x) = x^3$$

show that

$$\frac{g(2)-g(-1)}{f(2)-f(-1)} = 3$$

but that there is no point u in $(-1, 2)$ such that

$$\frac{g'(u)}{f'(u)} = 3.$$

Why does this not contradict the Cauchy mean value theorem?

6.4. Applications of derivatives

Higher order derivatives. If a function f is differentiable in a given interval, then we may obtain another function f', its derivative. If the function f' is itself differentiable in this interval, then we may obtain its derivative which we shall write f''. We call f'' the second derivative of the function f. As long as subsequent derivatives are themselves differentiable we may repeat this process and obtain the nth derivative of f, which we shall write $f^{(n)}$.

For example, if $f(x) = x^4$, then $f'(x) = 4x^3, f''(x) = 12x^2, f^{(3)}(x) = 24x$, $f^{(4)}(x) = 24$, $f^{(5)}(x) = 0$, and so on.

It is not always possible to obtain higher order derivatives. A function may be differentiable at a without being twice differentiable at a. For example, we saw in section 6.2 that the function f defined by

$$\begin{cases} f(x) = x^2 \sin \dfrac{1}{x} & \text{if} \quad x \neq 0, \\ f(0) = 0, \end{cases}$$

is differentiable everywhere, but that its derivative f' is discontinuous at zero. Therefore f' is not differentiable at zero and $f''(0)$ does not exist.

Maxima and minima

DEFINITION. *If the function f is defined in some open interval containing the point a and if there is a real number $\delta > 0$, such that $f(a) > f(x)$ whenever $0 < |x-a| < \delta$, then f is said to have a* (local) maximum *at a.*

DEFINITION. *If the function f is defined in some open interval containing the point a and if there is a real number $\delta > 0$, such that $f(a) < f(x)$ whenever $0 < |x-a| < \delta$, then f is said to have a* (local) minimum *at a.*

THEOREM 6.4.1. *If the function f is differentiable at the point a and has a maximum or minimum at a, then $f'(a) = 0$.*

Proof. This theorem follows immediately from Theorem 6.2.1, since the conditions on the function f here are more exacting than those of Theorem 6.2.1.

The following results provide criteria for testing whether a function has maxima or minima.

THEOREM 6.4.2. (i) *The function f is differentiable in the open interval (a, b). f is an increasing function in (a, b) if and only if its derivative $f'(x) \geqslant 0$ everywhere in (a, b).*

(ii) *The function f is differentiable in the open interval (a, b). f is a decreasing function in (a, b) if and only if its derivative $f'(x) \leqslant 0$ everywhere in (a, b).*

Proof. (i) Suppose, first, that $f'(x) \geqslant 0$ everywhere in (a, b). If $a < u < v < b$, then f satisfies the conditions of the first mean value theorem in $[u, v]$. Therefore there is a point w in (u, v) such that

$$\frac{f(v)-f(u)}{v-u} = f'(w).$$

But $f'(w) \geqslant 0$, so that $f(v) \geqslant f(u)$. Therefore f is increasing.

Suppose now that f is increasing in (a, b). If u is any point in (a, b) then

$$f'(u) = \lim_{h \to 0+} \left[\frac{f(u+h)-f(u)}{h}\right].$$

Since f is increasing, $f(u+h)-f(u)/h \geqslant 0$. Therefore $f'(u) \geqslant 0$.

(ii) May be proved similarly. □

THEOREM 6.4.3. *If there is a real number $\delta > 0$ such that the function f is continuous in $(a-\delta, a+\delta)$, $f'(x) > 0$ in $(a-\delta, a)$, and $f'(x) < 0$ in $(a, a+\delta)$, then f has a maximum at the point a.*

Proof. If u is a point in $(a-\delta, a)$, then f satisfies the conditions of the first mean value theorem in $[u, a]$. Therefore there is a point t in (u, a) such that

$$\frac{f(a)-f(u)}{a-u} = f'(t) > 0.$$

Therefore $f(a) > f(u)$.

If v is a point in $(a, a+\delta)$, then f satisfies the conditions of the first mean value theorem in $[a, v]$. Therefore there is a point w in (a, v) such that

$$\frac{f(v)-f(a)}{v-a} = f'(w) < 0.$$

Therefore $f(v) < f(a)$. Hence f has a maximum at a. □

THEOREM 6.4.4. *If there is a real number $\delta > 0$ such that the function f is continuous in $(a-\delta, a+\delta)$, $f'(x) < 0$ in $(a-\delta, a)$, and $f'(x) > 0$ in $(a, a+\delta)$, then f has a minimum at the point a.*

The proof of this theorem is similar to that of Theorem 6.4.3.

THEOREM 6.4.5. *If the function f is twice differentiable at the point a and if*

(i) $f'(a) = 0, f''(a) < 0$, *then f has a maximum at a.*

(ii) $f'(a) = 0, f''(a) > 0$, *then f has a minimum at a.*

Proof. (i) Since $f'(a) = 0$,

$$f''(a) = \lim_{h\to 0}\left[\frac{f'(a+h)}{h}\right].$$

Thus, since $f''(a) < 0$,

$$\frac{f'(a+h)}{h} < 0 \quad \text{for all sufficiently small } h.$$

Therefore there is a real number $\delta > 0$ such that

$$f'(a+h) > 0 \quad \text{whenever} \quad -\delta < h < 0,$$

and

$$f'(a+h) < 0 \quad \text{whenever} \quad 0 < h < \delta.$$

Thus f satisfies the conditions of Theorem 6.4.3 and therefore has a maximum at a.

(ii) Follows similarly from Theorem 6.4.4. □

We note that Theorem 6.4.5 does not impose *necessary* conditions for a function to have a maximum or a minimum. For example, if $f(x) = x^4$, f has a minimum at zero even though f does not satisfy the conditions of Theorem 6.4.5(ii), since $f'(0) = 0$ and $f''(0) = 0$. If $g(x) = |x|$, then g has a minimum at zero even though $g'(0)$ and $g''(0)$ do not even exist.

Anti-derivatives

THEOREM 6.4.6. *If the function f is differentiable in (a, b) and if $f'(x) = 0$ everywhere in (a, b), then f is constant in (a, b).*

Proof. Let u and v be points of (a, b) with $u < v$. f satisfies the conditions of the mean value theorem in the interval $[u, v]$. Thus there is a point w in $[u, v]$ such that

$$\frac{f(v)-f(u)}{v-u} = f'(w) = 0.$$

Therefore $f(u) = f(v)$. □

COROLLARY. *If f and g are differentiable functions in the interval (a, b) and if $f'(x) = g'(x)$ everywhere in (a, b) then f and g differ by a constant.*

Proof. Define the function h by $h(x) = f(x)-g(x)$. Then $h'(x) = f'(x)-g'(x) = 0$ everywhere in (a, b). Therefore h is constant by Theorem 6.4.6. □

In view of the fundamental theorem of calculus, which will be stated and proved in Chapter 8, it is very useful, when given a function f, to be able to find a function F whose derivative is equal to f.

DEFINITION. *Given a function f, if F is another function whose derivative F' is equal to f, then F is called an* anti-derivative *of f.*

(Many people use the term indefinite integral instead of anti-derivative. The latter term, however, is perhaps to be preferred since it is less confusing and does not prematurely anticipate the fundamental theorem of calculus.)

If f has an anti-derivative F, it has an infinite number of anti-derivatives. We deduce from Theorem 6.4.6 and its corollary that the set of all antiderivatives is

$$\{G : G(x) = F(x)+c, \quad c \text{ is any real number}\}.$$

L'HOPITAL'S RULE. *Derivatives are often used to calculate the value of indeterminate forms, i.e. expressions of the type*

$$\lim_{h\to 0}\left[\frac{f(x)}{g(x)}\right], \quad f(a) = g(a) = 0.$$

The method used is known as L'Hopital's rule. We prove it in two forms, imposing different conditions on the functions f and g.

THEOREM 6.4.7 (first form of L'Hopital's rule). *f and g are functions defined in some open interval containing a and $f(a) = g(a) = 0$. If f and g are both differentiable at a and if $g'(a) \neq 0$, then*

$$\lim_{h\to 0}\left[\frac{f(x)}{g(x)}\right] = \frac{f'(a)}{g'(a)}.$$

Proof.

$$\begin{aligned}\lim_{x\to a}\left(\frac{f(x)}{g(x)}\right) &= \lim_{x\to a}\left(\frac{f(x)-f(a)}{g(x)-g(a)}\right)\\ &= \lim_{x\to a}\left[\frac{f(x)-f(a)}{x} \div \frac{g(x)-g(a)}{x}\right]\\ &= \lim_{x\to a}\left[\frac{f(x)-f(a)}{x}\right] \lim_{x\to a}\left[\frac{g(x)-g(a)}{x}\right]\\ &= \frac{f'(a)}{g'(a)}. \quad \square\end{aligned}$$

Thus, for example, $$\lim_{x\to\pi/2}\left[\frac{1-\sin x}{x-\pi/2}\right]=\frac{-\cos\pi/2}{1}=0$$

or $$\lim_{x\to 1}\left[\frac{\sqrt{x}-1}{x^3-1}\right]=\frac{1}{2\sqrt{1}}\div 3(1)^2=\frac{1}{6}$$

THEOREM 6.4.8 (second form of L'Hopital's rule). *The functions f and g are continuous at a and $f(a)=g(a)=0$. If*

$$\lim_{x\to a}\left[\frac{f'(x)}{g'(x)}\right]$$

exists, then so does

$$\lim_{x\to a}\left[\frac{f(x)}{g(x)}\right],$$

and the two limits are equal.

Proof. Since

$$\lim_{x\to a}\left[\frac{f'(x)}{g'(x)}\right]$$

exists, f' and g' must both exist in some interval $(a-\delta, a+\delta)$, where δ is a real number greater than zero. If u is a point of $(a, a+\delta)$, then f and g satisfy the conditions of Cauchy's mean value theorem in the interval $[a, u]$. Thus there is a point t in (a, u) such that

$$\frac{f(u)-f(a)}{g(u)-g(a)}=\frac{f'(t)}{g'(t)}$$

or $$\frac{f(u)}{g(u)}=\frac{f'(t)}{g'(t)}.$$

$t\to a$ as $u\to a$, since $t\in(a, u)$.

Therefore $$\lim_{x\to a+}\left[\frac{f(x)}{g(x)}\right]=\lim_{x\to a+}\left[\frac{f'(x)}{g'(x)}\right].$$

If v is a point of $(a-\delta, a)$, then f and g satisfy the conditions of Cauchy's mean value theorem in the interval $[v, a]$. Thus there is a point w in (v, a) such that

$$\frac{f(a)-f(v)}{g(a)-g(v)} = \frac{f'(w)}{g'(w)}$$

or

$$\frac{f(v)}{g(v)} = \frac{f'(w)}{g'(w)}.$$

$w \to a$ as $v \to a$, since $w \in (v, a)$.

Therefore $$\lim_{x\to a-}\left[\frac{f(x)}{g(x)}\right] = \lim_{x\to a-}\left[\frac{f'(x)}{g'(x)}\right].$$

But

$$\lim_{x\to a}\left[\frac{f'(x)}{g'(x)}\right]$$

exists; hence, by Theorem 5.6.1,

$$\lim_{x\to a}\left[\frac{f(x)}{g(x)}\right]$$

exists and the two limits are equal. □

Corollary 1. *The functions f' and g' are continuous at a and $f(a) = g(a) = f'(a) = g'(a) = 0$. If*

$$\lim_{x\to a}\left[\frac{f''(x)}{g''(x)}\right]$$

exists, then so do

$$\lim_{x\to a}\left[\frac{f(x)}{g(x)}\right] \quad \textit{and} \quad \lim_{x\to a}\left[\frac{f'(x)}{g'(x)}\right]$$

and the three limits are equal.

Proof. We apply Theorem 6.4.8, first to the functions g' and t' and then to the functions f and g. □

COROLLARY 2. *The functions* $f^{(n-1)}$ *and* $g^{(n-1)}$ *are continuous at a and f and g and their first* $n-1$ *derivatives are all zero at a. If*

$$\lim_{x\to a}\left[\frac{f^{(n)}(x)}{g^{(n)}(x)}\right]$$

exists, then so does

$$\lim\left[\frac{f(x)}{g(x)}\right]$$

and the two limits are equal.

Proof. We apply Theorem 6.4.8 $n-1$ times. □

For example, $$\lim_{x\to 0}\left[\frac{x-\sin x}{x^2}\right] = \lim_{x\to 0}\left[\frac{1-\cos x}{2x}\right] = \lim_{x\to 0}\left[\frac{\sin x}{2}\right] = 0.$$

Exercises 6.4

1. The function f is defined by

$$\begin{cases} f(x) = x^3 & \text{if } x \geqslant 0, \\ f(x) = 0 & \text{if } x < 0. \end{cases}$$

How many times is f differentiable at zero?

2. The function f is defined by

$$\begin{cases} f(x) = x\sin\dfrac{1}{x} & \text{if } x = 0, \\ f(0) = 0. \end{cases}$$

Show that f is continuous, but not differentiable, at zero.

3. The function f is defined by

$$\begin{cases} f(x) = x^5\sin\dfrac{1}{x} & \text{if } x \neq 0, \\ f(0) = 0. \end{cases}$$

Show that f is differentiable twice at zero, but is not differentiable three times there.

4. The function f is defined by

$$\begin{cases} f(x) = x^2 & \text{if} \quad x \text{ is a rational,} \\ f(x) = 0 & \text{if} \quad x \text{ is an irrational.} \end{cases}$$

Show that there is one point at which f is differentiable. How many times is f differentiable at this point?

5. Prove that if $f'(x) > 0$ in the open interval (a, b), then f is strictly increasing in this interval. Compare this result with Theorem 6.4.2. Show that the converse of this result, that if f is strictly increasing in (a, b), then $f'(x) > 0$ in (a, b), is not true.

6. Find the minima of f in each of the following cases. In each case say whether f satisfies (a) the conditions of Theorem 6.4.4, (b) the conditions of Theorem 6.4.5 (f may satisfy the conditions of both theorems or of neither theorem).

(i) $f(x) = x^3 - x$.

(ii) $f(x) = x^3$.

(iii) $f(x) = x^6$.

(iv) $\begin{cases} f(x) = x^2 & \text{if} \quad x \geqslant 0, \\ f(x) = x^4 & \text{if} \quad x < 0. \end{cases}$

(v) $\begin{cases} f(x) = x^2 & \text{if} \quad x \text{ is a rational,} \\ f(x) = x^4 & \text{if} \quad x \text{ is an irrational.} \end{cases}$

7. The function f is defined by

$$\begin{cases} f(x) = x^2 & \text{if} \quad x \text{ is a rational,} \\ f(x) = x^4 & \text{if} \quad x \text{ is an irrational} \end{cases}$$

Show that $f'(0) = 1$. If $g(x) = \sin x$ find the value of

$$\lim_{x \to 0} \left[\frac{f(x)}{g(x)}\right].$$

Which form of L'Hopital's rule must be used and why?

8. Explain how Theorem 6.4.1 differs from Theorem 6.2.1.

9. Prove Theorems 6.4.2 (ii), 6.4.4, and 6.4.5 (ii).

10. Use L'Hopital's rule to evaluate the following limits:

(i) $\lim_{x \to 1} \left[\frac{x^3 - 1}{x^5 - 1}\right]$.

(ii) $\lim_{x \to \pi} \left(\frac{\sin 3x}{\sin 2x}\right)$.

(iii) $\lim_{x \to 0} \left(\frac{x - \sin x}{x^3}\right)$.

(iv) $\lim_{x \to 0} \left[\frac{\sqrt{(1-x)} + \frac{1}{2}x - 1}{x^2}\right]$.

(v) $\lim_{x \to 0} \left[\frac{\cos x - 1}{x \sin x}\right]$.

(vi) $\lim_{x \to 0} \left[\frac{\sqrt{(1+x)} + \sqrt{(1-x)} - 2}{x^2}\right]$.

6.5. Taylor series

Suppose we know the value of some function f at a, but not at $a+h$. If h is fairly small and we wish to guess the value of $f(a+h)$, then, rather than guess at random, we may reasonably guess the number $f(a)$ on the grounds that the function f may not have changed much between a and $a+h$. That is we guess

$$f(a+h) \simeq f(a).$$

If $f'(x)$ is small throughout the interval $(a, a+h)$, then Theorem 6.3.1 tells us that the approximation is close.

Now suppose that, in addition to knowing $f(a)$ we also know $f'(a)$. We may now guess a value for $f(a+h)$ on the hypothesis that the rate of change of f in $(a, a+h)$ is constant. In this way we obtain

$$f(a+h) \simeq f(a)+hf'(a).$$

Now suppose that we know, in addition, $f''(a)$. This will tell us whether the rate of change of f is increasing or decreasing at a and how quickly, thus causing us to adjust our guess for $f(a+h)$. On the hypothesis that the rate of change of f is changing at a uniform rate, the average rate of change of f in $(a, a+h)$ is

$$f'(a)+\frac{h}{2}f''(a).$$

Thus our revised guess for $f(a+h)$ is

$$\begin{aligned} f(a+h) &\simeq f(a)+h\left(f'(a)+\frac{h}{2}f''(a)\right) \\ &= f(a)+hf'(a)+\frac{h^2}{2}f''(a). \end{aligned}$$

We may continue giving ourselves the values of higher order derivatives of f at a; we see that, unless the function f is very badly behaved, the more derivatives we know the better is our estimate of $f(a+h)$. Taylor's theorem deals with this situation.

THEOREM 6.5.1 (Taylor's theorem). *If f is a function whose $(n-1)$th derivative is continuous in the closed interval $[a, a+h]$ and whose nth derivative exists in $(a, a+h)$, then*

$$f(a+h) = f(a)+hf'(a)+\frac{h^2}{2!}f''(a)$$

$$+\frac{h^3}{3!}f'''(a)+\ldots+\frac{h^{n-1}}{(n-1)!}f^{(n-1)}(a)+R_n,$$

where R_n may have either *of the following forms:*

(i) $R_n = (h^n/n!)f^{(n)}(a+\theta h)$ *for some θ satisfying* $0 < \theta < 1$
(Lagrange's form of remainder);

(ii) $R_n = \{h^n(1-\phi)^{n-1}/(n-1)!\}f^{(n)}(a+\phi h)$ *for some ϕ satisfying* $0 < \phi < 1$
(Cauchy's form of remainder).

Proof. (i) Let

$$F(x) = f(x)+(a+h-x)f'(x)$$

$$+\frac{(a+h-x)^2}{2!}f''(x)+\ldots+\frac{(a+h-x)^{n-1}}{(n-1)!}f^{(n-1)}(x)$$

and let $$G(x) = (a+h-x)^n.$$

The reader is left to verify that

$$F'(x) = \frac{(a+h-x)^{n-1}}{(n-1)!}f^{(n)}(x), \quad G'(x) = -n(a+h-x)^{n-1}.$$

F and G satisfy the conditions of Cauchy's mean value theorem in the interval $[a, a+h]$. Therefore there is a real number θ in $(0, 1)$ such that

$$\frac{F(a+h)-F(a)}{G(a+h)-G(a)} = \frac{F'(a+\theta h)}{G'(a+\theta h)}.$$

Thus

$$\frac{f(a+h)-\left\{f(a)+hf'(a)+\frac{h^2}{2!}f''(a)+\ldots+\frac{h^{n-1}}{(n-1)!}f^{(n-1)}(a)\right\}}{-h^n}$$

$$=\frac{-f^{(n)}(a+\theta h)}{n!}$$

or

$$f(a+h)=f(a)+hf'(a)+\frac{h^2}{2!}f''(a)+\ldots+$$

$$+\frac{h^{n-1}}{(n-1)!}f^{(n-1)}(a)+\frac{h^n}{n!}f^{(n)}(a+\theta h).$$

(ii) Define F as in (i) but let $G(x) = a+h-x$. Then $G'(x) = -1$. F and G satisfy the conditions of the Cauchy mean value theorem in the interval $[a, a+h]$. Therefore there is a real number ϕ in $(0, 1)$ such that

$$\frac{F(a+h)-F(a)}{G(a+h)-G(a)}=\frac{F'(a+\phi h)}{G'(a+\phi h)}.$$

Thus

$$\frac{f(a+h)-\left\{f(a)+hf'(a)+\frac{h^2}{2!}f''(a)+\ldots+\frac{h^{n-1}}{(n-1)!}f^{(n-1)}(a)\right\}}{-h}$$

$$=\frac{h^{n-1}(1-\phi)^{n-1}f^{(n)}(a+\phi h)}{-(n-1)!}$$

or

$$f(a+h)=f(a)+hf'(a)+\frac{h^2}{2!}f''(a)+\ldots+$$

$$+\frac{h^{n-1}}{(n-1)!}f^{(n-1)}(a)+\frac{h^n(1-\phi)^{n-1}}{(n-1)!}f^{(n)}(a+\phi h).\ \square$$

We have tacitly assumed in the statement and proof of this theorem that $h > 0$, since the intervals $[a, a+h]$ and $(a, a+h)$ have been mentioned. The modifications, in both statement and proof, which are necessary in order that negative values of h may be included are, however, trivial. Hence the theorem will be regarded as true for all values of h.

Taylor series. In view of Taylor's theorem we make the following definition.

DEFINITION. *The* Taylor series *for the function f at the point a is*

$$\sum_{n=0}^{\infty} \frac{(x-a)^n}{n!} f^{(n)}(a),$$

where $0!$ *is defined to be equal to* 1.

If S_n is the nth partial sum of the Taylor series for f, then Taylor's theorem gives

$$S_n = f(x) - R_n.$$

Thus the function f will be equal to its Taylor series if and only if $\lim R_n = 0$. There are three possibilities for the behaviour of $\{R_n\}$ which are of interest.

(i) $\{R_n\}$ is divergent. In this case the Taylor series is divergent.

(ii) $\{R_n\}$ is convergent, but $\lim R_n \neq 0$. In this case the Taylor series is convergent, but its sum to infinity is not equal to f.

(iii) $\lim R_n = 0$. In this case f is equal to the sum to infinity of its Taylor series.

Thus, in order to prove that a function is equal to its Taylor series, it is not sufficient to show that the Taylor series is convergent. We have rather to show that $\lim R_n = 0$. In order to do this for a particular function we may use either the Lagrange form of remainder or the Cauchy form. If possible we use the Lagrange form because it is simpler. In this chapter we give examples of the use only of the Lagrange form: the use of the Cauchy form will be exemplified in sections 7.4 and 7.6.

DEFINITION. *The* Maclaurin series *for the function f is the Taylor series for f at the point zero. Thus the Maclaurin series is*

$$\sum_{n=0}^{\infty} \frac{x^n}{n!} f^{(n)}(a).$$

We conclude this section by giving the Maclaurin series for $\sin x$ and for $\cos x$ and an example of a Taylor series.

THEOREM 6.5.2. (i) *The Maclaurin series for* $\sin x$ *is*

$$\sum_{n=0}^{\infty} \frac{(-1)^n x^{2n+1}}{(2n+1)!}$$

and $\sin x$ *is equal to its Maclaurin series for all values of* x.

(ii) *The Maclaurin series for* $\cos x$ *is*

$$\sum_{n=0}^{\infty} \frac{(-1)^n x^{2n}}{(2n)!}$$

and $\cos x$ *is equal to its Maclaurin series for all values of* x.

Proof. (i) Let $f(x) = \sin x$. Then

$$f^{(4n)}(x) = \sin x, \quad f^{(4n+1)}(x) = \cos x, \quad f^{(4n+2)}(x) = -\sin x,$$
$$f^{(4n+3)}(x) = -\cos x.$$

Thus

$$f^{(4n)}(0) = 0, \quad f^{(4n+1)}(0) = 1, \quad f^{(4n+2)}(0) = 0, \quad f^{(4n+3)}(0) = -1.$$

Hence the Maclaurin series for $\sin x$ is

$$\sum_{n=0}^{\infty} \frac{(-1)^n x^{2n+1}}{(2n+1)!}.$$

We now have to show that $\lim R_n = 0$. We have $|f^{(n)}(x)| \leqslant 1$ for all natural numbers n and for all real numbers x. Therefore

$$|R_n| \leqslant \frac{|x|^n}{n!}.$$

But $\lim |x|^n/n! = 0$ by Theorem 3.6.4. Therefore $\lim R_n = 0$.

The proof of (ii) is similar and is left to the reader. □

We now find the Taylor series at the point 2 for the function $f(x) = 1/x$ and show that $1/x$ is equal to the sum to infinity of its Taylor series when $2 < x < 4$.

$$f^{(n)}(x) = \frac{(-1)^n n!}{x^{n+1}}.$$

Thus the Taylor series is

$$\sum_{n=0}^{\infty} \frac{(-1)^n (x-2)^n}{2^{n+1}}.$$

When $2 < x < 4$,

$$R_n = \frac{(-1)^n (x-2)^n}{u^{n+1}} \quad \text{for some } u \text{ satisfying } 2 < u < x.$$

But

$$\lim_{n\to\infty} \left(\frac{x-2}{u}\right)^n = 0 \quad \text{since} \quad 0 < \frac{x-2}{2} < 1.$$

Therefore $\lim R_n = 0$. Hence $1/x$ is equal to the sum to infinity of its Taylor series. (In fact, we can also prove that $\lim R_n = 0$ when $0 < x < 2$. This, however, requires the use of the Cauchy form of remainder.)

Exercises 6.5

1. Prove Theorem 6.5.2 (ii).

2. Find the Maclaurin series for the function $f(x) = 1/(1+x)$. By using the Lagrange form of remainder show that $\lim R_n = 0$, when $0 < x < 1$.

3. Find the Maclaurin series for the function $f(x) = 1/(1-2x)^2$. By using the Lagrange form of remainder show that $\lim R_n = 0$, when $-\frac{1}{2} < x < 0$.

4. Obtain the Taylor series for $\cos x$ at the point $\pi/6$ (assume that $\cos(\pi/6) = \sqrt{3}/2$ and $\sin(\pi/6) = \frac{1}{2}$). Show that $\lim R_n = 0$, by using the Lagrange form of remainder.

Hence evaluate $\cos 32°$, correct to two places of decimals. (Take $\pi = 3.142$ and $\sqrt{3} = 1.732$).

5. Use the Maclaurin series for $\sin x$ to show that, for $0 < x \leqslant \pi/2$, $\sin x < x$, and $\sin x > x - x/6$.

Deduce that $0.0174 < \sin 1° < 0.0175$, by assuming that $3.141 < \pi < 3.142$.

Exercises for Chapter 6

1. Sketch the Cartesian graph of the function

$$f(x) = 2x^3 - 16x + 15.$$

Find the values of x for which the function has a maximum or a minimum. Mark on the graph the points for which $x = 1$ and $x = 2$.

If $\{a_n\}$ is defined by the recurrence relation

$$16a_{n+1} = 2a_n^3 + 15$$

with $a_n = 1$, show, by induction on n, that

$$\text{(i) } a_{n+1} > a_n, \quad \text{(ii) } a_n < 2,$$

for all natural numbers n.

Deduce that $\{a_n\}$ converges to one of the roots of the equation $2x^3 - 16x + 15 = 0$ and indicate on the sketch which root. (See section 3.5.)

2. Explain briefly what you understand by saying that a function $f(x)$ is continuous at $x = a$.

Discuss the continuity and differentiability of the following functions at $x = 0$:

(i) $f(x) = 1 - x - \lim_{n \to \infty} (\cos^{2n+1} \pi x)$,

(ii) $g(x) = x \sin \dfrac{1}{x} \; (x \neq 0), \quad g(0) = 0$,

(iii) $h(x) = e^{1/x} \; (x \neq 0), \quad h(0) = 0.$ [B.Ed.]

3. Show that a function which is differentiable at $x = a$ is also continuous at $x = a$, and give an example to show that the converse of this result is false.

Sketch the graph of the function f given by

$$f(x) = |x^3| \quad \text{for all real } x.$$

Show that f is differentiable everywhere (including at $x = 0$), and give an expression for its derivative f'. State, giving reasons, whether the second and third derivatives f'' and f''' exist at $x = 0$. [T.C.]

4. State Rolle's Theorem.

The functions f and g are continuous in the closed interval $[a, b]$ and differentiable in the open interval (a, b). By considering the function h given by

$$h(x) = \{f(b)-f(a)\}\,g(x)-\{g(b)-g(a)\}\,f(x),$$

prove that there is a ξ in the interval (a, b) such that

$$\{f(b)-f(a)\}\,g'(\xi) = \{g(b)-g(a)\}\,f'(\xi).$$

A curve is given parametrically by $x = f(t)$, $y = g(t)$. Show that the gradient of curve at the point with parameter t (if it exists) is given by $g'(t)/f'(t)$. Explain the result about $f'(\xi)$ and $g'(\xi)$ geometrically in terms of this curve.

By constructing another suitable function and applying Rolle's theorem, show that there is an η in (a, b) such that

$$\{f(\eta)-f(a)\}\,g'(\eta) = \{g(b)-g(\eta)\}\,f'(\eta).$$

Describe this result also in terms of the curve. [T.C.]

5. State carefully and prove the mean value theorem concerning the equation

$$f(b)-f(a) = (b-a)f'(\xi).$$

If Rolle's theorem is used in your proof, this theorem should be stated.

Why can the mean value theorem not be applied to either of the following?

(a) $f(x) = |1-x|$ in the closed interval $[0, 3]$;

(b) $f(x) = 1/x^2$ in the closed interval $[-\frac{1}{2}, \frac{1}{2}]$.

Find the appropriate value of ξ when the mean value theorem is applied to the general quadratic function

$$f(x) = lx^2+mx+n$$

in the closed interval $[a, b]$. [T.C.]

6. Define the continuity of a function $f(x)$ at $x = a$. Prove that if $f(x)$ is differentiable at $x = a$, then it is also continuous there. Under what conditions is it true that

$$f(a+h) = f(a)+hf'(a)+\ldots+\frac{h^{n-1}}{(n-1)!}f^{n-1}(a)+\frac{h^n}{n!}f^{(n)}(a+\theta h)?$$

Show that the function

$$f(x) = \begin{cases} \dfrac{1}{x} & (0 < x \leqslant 1), \\[2mm] \dfrac{(x-1)^5+1}{x} & (x > 1) \end{cases}$$

is continuous at every point of its domain and has continuous derivatives $f^n(x)$ at every point of its domain for $n = 1, 2, \ldots, k$, where k should be found. State any results about continuous functions which you use. [T.C.]

7. State Taylor's theorem for a function in the closed interval $[a, b]$. Show that

$$\sin\beta = \sin\alpha+(\beta-\alpha)\cos\alpha-\frac{(\beta-\alpha)^2}{2!}\sin\alpha-\frac{(\beta-\alpha)^3}{3!}\cos\xi,$$

where $\alpha < \xi < \beta$.

Use this result to evaluate sin 51°, given that $\sin 45° = 1/\sqrt{2}$ and show that your answer is accurate to three decimal places. [T.C.]

8. By applying Rolle's theorem to the function

$$F(x) = f(x)-f(a)-\left\{\frac{x-a}{b-a}\right\}\{f(b)-f(a)\},$$

where $f(x)$ is continuous on the closed interval $[a, b]$ and differentiable on the open interval (a, b), prove the mean value theorem for the function $f(x)$.

(a) Calculate to two decimal places a positive value of Θ for which the tangent at point $(1+\Theta, (1+\Theta)^3)$ on the curve $y = x^3$ is parallel to the chord joining the points (1, 1), (2, 8) on the curve.

(b) Use the mean value theorem to prove l'Hopital's rule, namely that if f and g are functions of x which have continuous derivatives in an interval containing $x = a$, and if $f(a) = g(a) = 0$, $g'(a) \neq 0$, then $\lim_{x\to a} f(x)/g(x) = f'(a)/g'(a)$. [T.C.]

9. Define the derivative of $f(x)$ at $x = a$ and explain why it is necessary that $f(x)$ be continuous at $x = a$ for this derivative to exist.

Let

$$f(x) = \begin{cases} x^2\sin(1/x) & \text{when } x \neq 0, \\ 0 & \text{when } x = 0. \end{cases}$$

Prove that $f'(0) = 0$ and that $f'(x)$ is discontinuous at $x = 0$. [B.Ed.]

10. Show that if $f(x)$ is continuous on the closed interval $a \leqslant x \leqslant b$ and is differentiable in the open interval $a < x < b$, then the function

$$g(x) = f(x)-f(a)-\frac{(x-a)}{(b-a)}\{f(b)-f(a)\}$$

satisfies the conditions of Rolle's theorem. Hence or otherwise prove that

$$f(b)-f(a) = (b-a)f'(u)$$

for at least one u in the open interval $a < u < b$.

By suitably choosing f, or otherwise, prove that

$$3 \sin 3x + 2 \sin 2x + \sin x = 8/\pi$$

for at least one x in the interval $0 < x < \frac{1}{2}\pi$. [B.Ed.]

11. Prove that at a local maximum ξ of a function f inside an interval, in which f is differentiable, $f'(\xi) = 0$.

The function f is differentiable in $[0, 1]$, $f'(0) > 0$, $f'(1) < 0$. Prove that there is a number ξ, $0 < \xi < 1$, such that $f'(\xi) = 0$. If $f'(\alpha) = a$, $f'(\beta) = b$, $0 < \alpha < \beta < 1$, and c lies between a and b, by considering $f(x) - cx$, prove that there is a γ, $\alpha < \gamma < \beta$, such that $f'(\gamma) = c$. If g has a positive derivative in $(0, 1]$, and is defined in $[-1, 0)$ by $g(x) = g(-x)$, prove that it can be defined at 0 so as to make it continuous there, and that, if it is differentiable at 0, then $g'(0) = 0$. [B.Ed.]

12. The Taylor expansions of the functions $\cos 2x$ and x^{-2} in powers of $(x - \pi/4)$ are given by

$$\cos 2x = \sum_{n=0}^{\infty} a_n \left(x - \frac{\pi}{4}\right)^n, \quad x^{-2} = \sum_{n=0}^{\infty} b_n \left(x - \frac{\pi}{4}\right)^n.$$

Find a_n and b_n and state the range of values of x for which each expansion is valid.

Evaluate

$$\sum_{n=0}^{\infty} |a_n|,$$

and find the function whose Taylor expansion is

$$\sum_{n=0}^{\infty} |b_n| \left(x - \frac{\pi}{4}\right)^n.$$

[B.Sc.]

13. Draw the graphs of the following functions:

$$\sin 2x, \quad \sin^2 x, \quad \sin (x^2).$$

Determine, in the last case, the slope of the graph at each of its points of intersection with the x-axis.

$x_1, x_2, \ldots, x_n$ are the x-coordinates of the points in which the graph of $\sin(x^2)$ intersects the positive x-axis, written in ascending order of magnitude; prove that $x_n - x_{n-1} \to 0$ as $n \to \infty$. [B.Sc.]

14. Show from first principles that the derivative of $\cos x$ is $-\sin x$. Write down the derivative of

$$f(x) = \alpha \cos x - \cos 2x,$$

where α is a positive constant.

Sketch the graph of $f(x)$ from $x = 0$ to $x = \frac{1}{2}\pi$, distinguishing the cases $\alpha \geqslant 4$ and $0 < \alpha < 4$.

Find in each case the greatest and least values of $f(x)$ for $0 \leqslant x \leqslant \frac{1}{2}\pi$. [B.Sc.]

15. In Theorems 6.2.3 and 6.2.4 it was seen that derivatives *always* possess the intermediate value property characteristic of continuous functions, even though they are not always continuous.

Theorem 5.7.2 gives another property characteristic of continuous functions, but this property is not always true for derivatives.

Demonstrate this by considering the function f defined by

$$\begin{cases} f(x) = x\sqrt{|x|} \sin \dfrac{1}{x} & \text{if} \quad x \neq 0, \\ f(0) = 0, \end{cases}$$

and by showing that there is a closed interval in which f' exists without being bounded in that interval.

CHAPTER 7

SOME IMPORTANT FUNCTIONS AND EXPANSIONS

7.1. Power series

At the end of the previous chapter we saw how many common functions may be written in the form of a Taylor or Maclaurin series. We now take the opposite point of view; if we begin with a series of ascending powers of x, under what circumstances does this series define a function and what are the properties of a function so defined?

DEFINITION. *A* power series *is a series of the form*

$$\sum a_n x^n.$$

Maclaurin series are examples of power series.

Generally the convergence or divergence of a power series, when given the sequence $\{a_n\}$, will evidently depend upon the value of x. For example, the power series

$$\sum_{n=0}^{\infty} x^n$$

is convergent if $|x| < 1$ and divergent otherwise and the series

$$\sum_{n=0}^{\infty} 2^n x^n$$

is convergent if $|x| < \frac{1}{2}$ and divergent otherwise. The following theorem gives a rule for determining those values of x for which a power series converges.

THEOREM 7.1.1. *If* $\overline{\lim}\ |a|^{1/n} = 1/R$, *then the power series* $\sum a_n x^n$ *is absolutely convergent when* $|x| < R$ *and divergent when* $|x| > R$.

Proof. By Cauchy's test (Theorem 4.6.3), $\sum a_n x^n$ is absolutely convergent if

$$\overline{\lim}\ [|a_n x^n|^{1/n}] < 1,$$

i.e. if
$$|x|\ \overline{\lim}\ [|a_n|^{1/n}] < 1,$$

i.e. if
$$|x| < R.$$

So, too, $\sum a_n x^n$ is divergent if

$$\overline{\lim}\ [|a_n x^n|^{1/n}] > 1,$$

i.e. if
$$|x| > R. \ \square$$

In view of this theorem, $R = \overline{\lim}\ |a_n|^{1/n}$ is called the radius of convergence of the power series

$$\sum_{n=0}^{\infty} a_n x^n.$$

This theorem then tells us that a power series is absolutely convergent if $|x|$ is less than its radius of convergence and is divergent if $|x|$ is greater than its radius of convergence. If R is the radius of convergence and if $|x| = R$, then the series may either converge or diverge. The following examples illustrate this.

(i)
$$\sum_{n=0}^{\infty} \frac{1}{n^n} x^n.$$

Since
$$\overline{\lim} \left[\frac{1}{n^n}\right]^{1/n} = \overline{\lim}\ \frac{1}{n} = 0,$$

the radius of convergence of this series is ∞, i.e. the series converges for all values of x.

(ii)
$$\sum_{n=0}^{\infty} \frac{x^n}{n!} = \sum_{n=0}^{\infty} b_n.$$

For this series we use D'Alembert's ratio test.

$$\left|\frac{b_{n+1}}{b_n}\right| = \left|\frac{x}{n+1}\right| \quad \text{and} \quad \lim_{n\to\infty}\left|\frac{x}{n+1}\right| = 0$$

for all values of x. Therefore $\lim |b_{n+1}/b_n| = 0$ for all values of x, so that the radius of convergence is again ∞.

(iii) $$\sum_{n=0}^{\infty} \frac{x^n}{n^2}.$$

By Theorems 3.6.2 and 3.3.3,

$$\overline{\lim}\left(\frac{1}{n^2}\right)^{1/n} = 1.$$

Therefore the radius of convergence of this series is 1. When $x = 1$ the series is absolutely convergent, since

$$\sum_{n=1}^{\infty} \frac{1}{n^2}$$

is convergent.

(iv) $$\sum_{n=1}^{\infty} \frac{(2x)^n}{n}.$$

By Theorem 3.6.2,

$$\overline{\lim}\left[\frac{2^n}{n^2}\right]^{1/n} = \overline{\lim}\; 2\left(\frac{1}{n^2}\right)^{1/n} = 2.$$

Therefore the radius of convergence of this series is $\frac{1}{2}$. When $x = \frac{1}{2}$, the series becomes

$$\sum_{n=0}^{\infty} \frac{1}{n},$$

which is divergent. When $x = -\frac{1}{2}$, the series becomes

$$\sum_{n=0}^{\infty} \frac{(-1)^n}{n}$$

which is convergent.

(v) $$\sum_{n=0}^{\infty} n^n \cdot x^n.$$

Since

$$\overline{\lim}\, (n^n)^{1/n} = \overline{\lim}\, n = \infty,$$

the radius of convergence of this series is zero, i.e. the series converges only when x is zero.

Differentiation of a power series. If the function f is defined by the equation

$$f(x) = \sum_{n=0}^{N} a_n x^n,$$

then f is differentiable everywhere, the derivative being

$$f'(x) = \sum_{n=0}^{N} n a_n x^{n-1}.$$

If, however, f is defined as the sum to infinity of a power series by the equation

$$f(x) = \sum_{n=0}^{\infty} a_n x^n \quad \text{if} \quad |x| < R,$$

where R is the radius of convergence of the power series, is the function f still differentiable and may we assume that its derivative is given by

$$f'(x) = \sum_{n=0}^{\infty} n a_n x^{n-1}?$$

The answer is that we may and we prove this by means of the next three theorems. The proof of Theorem 7.1.4 is difficult, but the result itself is relatively easy to understand and is of fundamental importance.

THEOREM 7.1.2. *If the radius of convergence of the series*

$$\sum_{n=0}^{\infty} a_n x^n$$

is R and if the radius of convergence of the series

$$\sum_{n=0}^{\infty} na_n x^{n-1}$$

is S, then $R = S$.

Proof.

$$\begin{aligned} R &= \overline{\lim}\, |a_n|^{1/n}, \\ S &= \overline{\lim}\, |na_n|^{1/n} = \overline{\lim}\, [n^{1/n} |a_n|^{1/n}] \\ &= \lim (n^{1/n})\, \overline{\lim}\, |a_n|^{1/n} \\ &= 1 \cdot R = R. \quad \square \end{aligned}$$

THEOREM 7.1.3. $\left| \dfrac{(x+h)^n - x^n}{h} \right| \leqslant n[|x|+|h|]^{n-1}.$

Proof.

$$\left| \frac{(x+h)^n - x^n}{h} \right|$$

$$= \left| \frac{(x+h-x)[(x+h)^{n-1} + x(x+h)^{n-2} + \ldots + x^{n-1}]}{h} \right|$$

$$\leqslant |x+h|^{n-1} + |x+h|^{n-2}|x| + \ldots + |x|^{n-1}$$

$$\leqslant n(|x|+|h|)^{n-1}.$$

THEOREM 7.1.4. *If the function f is defined by*

$$f(x) = \sum_{n=0}^{\infty} a_n x^n$$

whose radius of convergence is R, then f is differentiable in the interval $(-R, R)$ and

$$f'(x) = \sum_{n=0}^{\infty} na_n x^{n-1}.$$

Proof. Suppose $|X| < R$. Choose Y such that $|X| < Y < R$. Since the radius of convergence of $\sum_{n=0}^{\infty} na_n x^{n-1}$ is R, $\sum_{n=0}^{\infty} na_n Y^{n-1}$ is absolutely convergent. Thus, given $\varepsilon > 0$, we may find a natural number N such that

$$\sum_{n=N+1}^{\infty} |na_n Y^{n-1}| < \frac{\varepsilon}{3}.$$

We define the function g by

$$g(x) = \sum_{n=0}^{N} a_n x^n.$$

Then g is differentiable and its derivative is

$$g'(x) = \sum_{n=0}^{N} na_n x^{n-1}.$$

Thus there is a real number δ satisfying $0 < \delta < Y - |X|$ and such that

$$\left| \frac{g(X+h)-g(X)}{h} - \sum_{n=0}^{N} na_n X^{n-1} \right| < \frac{\varepsilon}{3} \quad \text{whenever} \quad 0 < |h| < \delta.$$

$$\left| \frac{f(X+h)-f(X)}{h} - \frac{g(X+h)-g(X)}{h} \right|$$

$$= \left| \sum_{n=N+1}^{\infty} \frac{a_n[(X+h)^n - X^n]}{h} \right| \leqslant \sum_{n=N+1}^{\infty} \left| \frac{a_n[(X+h)^n - X^n]}{h} \right|$$

$$< \sum_{n=N+1}^{\infty} |na_n(|X|+|h|)^{n-1}| < \sum_{n=N+1}^{\infty} |na_n Y^{n-1}| < \frac{\varepsilon}{3}.$$

Also
$$\left| \sum_{n=0}^{N} na_n X^{n-1} - \sum_{n=0}^{\infty} na_n X^{n-1} \right|$$

$$\leqslant \sum_{n=N+1}^{\infty} |na_n X^{n-1}|$$

$$< \sum_{n=N+1}^{\infty} |na_n Y^{n-1}| < \frac{\varepsilon}{3}.$$

Thus
$$\left|\frac{f(X+h)-f(X)}{h}-\sum_{n=0}^{\infty} na_n X^{n-1}\right|$$
$$\leqslant\left|\frac{f(X+h)-f(X)}{h}-\frac{g(X+h)-g(X)}{h}\right|+\left|\frac{g(X+h)-g(X)}{h}-\sum_{n=0}^{N} na_n X^{n-1}\right|$$
$$+\left|\sum_{n=0}^{N} na_n X^{n-1}-\sum_{n=0}^{\infty} na_n X^{n-1}\right|$$
$$<\frac{\varepsilon}{3}+\frac{\varepsilon}{3}+\frac{\varepsilon}{3}=\varepsilon.$$

Therefore
$$f'(X)=\lim\left[\frac{f(X+h)-f(X)}{h}\right]=\sum_{n=0}^{\infty} na_n X^{n-1}. \square$$

THEOREM 7.1.5. *A power series with radius of convergence R defines a function f which is continuous in the interval* $(-R, R)$.

Proof. This result is an immediate consequence of Theorems 7.1.4 and 6.1.1. □

Uniqueness of the Maclaurin series. At the end of the previous chapter we discussed functions which may be differentiated as often as we please. Given such a function f, we defined its Maclaurin series to be
$$\sum_{n=0}^{\infty}\frac{f^{(n)}(0)}{n!}x^n.$$

If the remainder term R_n tends to zero as n tends to infinity when $|x|<R$, then we have
$$f(x)=\sum_{n=0}^{\infty}\frac{f^{(n)}(0)}{n!}x^n \quad \text{if} \quad |x|<R.$$

We now show that, if a function may be expressed as a power series, this expression is unique, i.e. we are never able to find two different power series, both of which give the function f.

THEOREM 7.1.6. *If $R > 0$ and if*

$$f(x) = \sum_{n=0}^{\infty} a_n x^n = \sum_{n=0}^{\infty} b_n x^n \quad \text{whenever} \quad |x| < R,$$

then $a_m = b_m$ for all m.

Proof. By Theorem 7.1.4, f is differentiable m times in the interval $(-R, R)$ and

$$f^{(m)}(x) = \sum_{n=0}^{\infty} n(n-1)(n-2)\ldots(n-m+1)a_n x^{n-m}.$$

Therefore $f^{(m)}(0) = m!\, a_m$. So too, $f^{(m)}(0) = m!\, b_m$. Therefore $a_m = b_m$. □

Exercises 7.1

1. For each of the following series find the radius of convergence R and say whether the series is convergent or divergent when $x = R$ and when $x = -R$.

(i) $\sum_{n=0}^{\infty} nx^n$.

(ii) $\sum_{n=0}^{\infty} n!\, x^n$.

(iii) $\sum_{n=0}^{\infty} \frac{1}{\sqrt{n}} \left(\frac{x}{3}\right)^n$.

(iv) $\sum_{n=0}^{\infty} \frac{(4x)^n}{n^3}$.

(v) $\sum_{n=0}^{\infty} (-x)^n \sin\frac{1}{n}$.

2. State the radius of convergence and the sum to infinity of the power series

$$\sum_{n=0}^{\infty} x^n.$$

Use Theorem 7.1.4 to deduce the power series expansion of

$$\frac{1}{(1-x)^2}$$

and state the radius of convergence of this series.

7.2. The exponential function

As we mentioned in the introductory chapter, many situations are appropriately described by a function whose rate of change is proportional to its size at any instant. We shall now construct such functions. In particular, we shall first construct a function f whose rate of change is equal to its size at any instant, so that

$$f'(x) = f(x).$$

We look for a power series of the form

$$\sum_{n=0}^{\infty} a_n x^n$$

which will give such a function. $f(0) = a_0$ and we suppose that $f(0) = 1$. By Theorem 7.1.4,

$$f'(x) = \sum_{n=0}^{\infty} n a_n x^{n-1},$$

so that $f'(0) = a_1$. But $f'(0) = f(0)$, so that $a_1 = 1$.

$$f''(x) = \sum_{n=0}^{\infty} n(n-1)a_n x^{n-2},$$

so that $f''(0) = 2a_2$. But $f''(0) = f(0) = 1$, so that $a_2 = \frac{1}{2}$.

Continuing in this way we find that $a_n = 1/n!$. This leads us to make the following definition.

Definition. *The exponential function is defined by the equation*

$$\exp x = \sum_{n=0}^{\infty} \frac{x^n}{n!}.$$

We note immediately that $\exp 0 = 1$.

THEOREM 7.2.1. *The radius of convergence of the series*

$$\sum_{n=0}^{\infty} \frac{x^n}{n!}$$

is ∞, *so that* $\exp x$ *is defined for all values of* x.

Proof. We have already proved this as example (ii) following Theorem 7.1.1. □

THEOREM 7.2.2. *If* $f(x) = \exp x$, *then* f *is differentiable everywhere and* $f'(x) = f(x) = \exp x$.

Proof. By Theorem 7.1.4 f is differentiable everywhere and

$$f'(x) = \sum_{n=0}^{\infty} \frac{nx^{n-1}}{n!} = \sum_{n=1}^{\infty} \frac{x^{n-1}}{(n-1)!} = \sum_{n=0}^{\infty} \frac{x^n}{n!} = \exp x.$$

We give two proofs of the following theorem which is the fundamental property of the exponential function.

THEOREM 7.2.3.

$$\exp(x+y) = \exp x \cdot \exp y.$$

First proof. Let $a = x+y$. Define the function f by

$$f(x) = \exp x \cdot \exp(a-x).$$

Then $$f'(x) = \exp x \cdot \exp(a-x) + \exp x \cdot -\exp(a-x) = 0.$$

Therefore, by Theorem 6.4.6, f is constant. Since

$$\exp 0 = 1, \qquad f(0) = \exp 0 \cdot \exp a = \exp a.$$

Therefore $$f(x) = \exp a \text{ for all } x.$$

Thus $$\exp x \cdot \exp(a-x) = \exp a$$

or $$\exp x \cdot \exp y = \exp(x+y).$$

Second proof. This proof illustrates the use of the Cauchy product of two series

$$\exp x = \sum_{n=0}^{\infty} \frac{x^n}{n!}, \qquad \exp y = \sum_{n=0}^{\infty} \frac{y^n}{n!}.$$

Both these series are absolutely convergent for all values of x and y. Therefore, by Theorem 4.9.2,

$$\exp x \cdot \exp y = \sum_{n=0}^{\infty} c_n,$$

where

where $$c_n = \frac{x^0 y^n}{0!\,n!} + \frac{x^1 y^{n-1}}{1!(n-1)!} + \frac{x^2 y^{n-2}}{2!(n-2)!} + \ldots + \frac{x^n y^0}{n!\,0!}.$$

But the binomial theorem gives

$$(x+y)^n = \frac{n!\,x^n}{n!} + \frac{n!\,x^{n-1}y}{(n-1)!\,1!} + \frac{n!\,x^{n-2}y^2}{(n-2)!\,2!} + \ldots + \frac{n!\,y^n}{n!}.$$

Therefore $$c_n = \frac{(x+y)^n}{n!}.$$

Thus $$\exp x \cdot \exp y = \sum_{n=0}^{\infty} \frac{(x+y)^n}{n!} = \exp(x+y).$$

THEOREM 7.2.4. *If* $f(x) = \exp x$, *then*

(i) $f(x) \neq 0$ *for all* x.

(ii) $f(x) > 0$ *for all* x.

(iii) f *is a strictly increasing function;* $f(x) > 1$ *when* $x > 0$ *and* $f(x) < 1$ *when* $x < 0$.

(iv) $f(x) \to \infty$ *as* $x \to \infty$; $f(x) \to 0$ *as* $x \to -\infty$.

Proof. (i) Suppose there is a real number u such that $f(u) = 0$. Then, by Theorem 7.2.3,

$$f(0) = f(u) \cdot f(-u) = 0 \cdot f(-u) = 0.$$

But $f(0) = 1$, giving a contradiction. Hence there is no real number u such that $f(u) = 0$.

(ii) According to Theorem 7.2.2 f is differentiable everywhere and is therefore continuous. Suppose there is a real number u such that $f(u) < 0$. $f(0) = 1 > 0$. Therefore, by the intermediate value theorem, there is a real number v between zero and u such that $f(v) = 0$. This contradicts (i). Hence there is no real number u such that $f(u) < 0$.

(iii) $f'(x) = f(x) > 0$ for all x. Therefore f is strictly increasing (see exercises 6.4, question 5). The rest follows.

(iv) From the definition of $\exp x$ it is evident that $f(x) > x$ for all $x > 0$. Therefore $f(x) \to \infty$ as $x \to \infty$. If $y > 0$, then

$$f(y) \cdot f(-y) = f(0) = 1 \quad \text{and} \quad f(y) \to \infty \quad \text{as} \quad y \to \infty.$$

Therefore, by Theorem 3.3.5, $f(-y) \to 0$ as $y \to \infty$. Put $x = -y$. Then $f(x) \to 0$ as $x \to -\infty$. □

We now show that $\exp x$ tends to infinity faster than any power of x.

THEOREM 7.2.5. *If m is a natural number and if the function f is defined by*

$$f(x) = \frac{\exp x}{x^m},$$

then $f(x) \to \infty$ as $x \to \infty$.

Proof. When $x > 0$, $\exp x = \sum_{n=0}^{\infty} \frac{x^n}{n!} > \frac{x^{m+1}}{(m+1)!}$.

Therefore
$$\frac{\exp x}{x^m} > \frac{x}{(m+1)!}.$$

But
$$\frac{x}{(m+1)!} \to \infty \quad \text{as} \quad x \to \infty.$$

Therefore
$$\frac{\exp x}{x^m} \to \infty \quad \text{as} \quad x \to \infty.$$

An alternative approach to the derivative of exp x. In view of the importance of Theorem 7.1.4 we have chosen to make use of it in developing the theory of the exponential function. However, since Theorem 7.1.4 is difficult to prove, we now offer an alternative development which does not involve the use of this theorem.

THEOREM 7.2.6.

$$\lim_{x\to 0}\left[\frac{\exp x-1}{x}\right]=1.$$

Proof.

$$\frac{\exp x-1}{x}=\frac{1+x+x^2/2!+x^3/3!+\ldots-1}{x}$$

$$=1+\frac{x}{2!}+\frac{x^2}{3!}+\frac{x^3}{4!}+\ldots$$

$$=1+x\sum_{n=0}^{\infty}\frac{x^n}{(n+2)!}.$$

$\sum_{n=0}^{\infty} 1/(n+2)!$ is convergent; so let S be its sum to infinity. Thus, when $|x|<1$,

$$\left|\sum_{n=0}^{\infty}\frac{x^n}{(n+2)!}\right|\leqslant\sum_{n=0}^{\infty}\left|\frac{x^n}{(n+2)!}\right|<\sum_{n=0}^{\infty}\frac{1}{(n+2)!}=S.$$

Given $\varepsilon>0$ we define $\delta=\min(\varepsilon/S,1)$. Then

$$\left|\frac{\exp x-1}{x}-1\right|=|x|\left|\sum_{n=0}^{\infty}\frac{x^n}{(n+2)!}\right|<\frac{\varepsilon}{S}S=\varepsilon \text{ whenever } 0<|x|<\delta.$$

Therefore

$$\lim_{x\to 0}\left[\frac{\exp x-1}{x}\right]=1. \ \square$$

We now give a new proof of Theorem 7.2.2. We use Theorem 7.2.6 and Theorem 7.2.3 (which we must now prove by means of the second proof), but we avoid using Theorem 7.1.4.

THEOREM 7.2.2. *If $f(x) = \exp x$, then f is differentiable everywhere and $f'(x) = f(x) = \exp x$.*

New proof.

$$\begin{aligned}
&\lim_{h\to 0}\left[\frac{f(x+h)-f(x)}{h}\right] \\
&= \lim_{h\to 0}\left[\frac{\exp(x+h)-\exp x}{h}\right] \\
&= \lim_{h\to 0}\left[\exp x\left(\frac{\exp h-1}{h}\right)\right] \\
&= \exp x\cdot\lim_{h\to 0}\left[\frac{\exp h-1}{h}\right] \\
&= \exp x.
\end{aligned}$$

Therefore f is differentiable and $f'(x) = \exp x$. □

The remaining results are proved as before.

e^x. We define the number e to be exp 1, so that

$$e = 1+1+\frac{1}{2!}+\frac{1}{3!}+\frac{1}{4!}+\dots.$$

THEOREM 7.2.7. *If p is an integer and q is a natural number,*

$$\text{(i)}\quad \exp p = e^p, \qquad \text{(ii)}\quad \exp p/q = e^{p/q}.$$

Proof. (i) By Theorem 7.2.3,

$$\exp n = \exp(1+1+\dots+1) = (\exp 1)^n = e^n.$$

(ii) By Theorem 7.2.3,

$$e^p = \exp p = \exp(p/q+p/q+\dots+p/q) = (\exp(p/q))^q$$

Therefore $$e^{p/q} = \exp(p/q).$$

Thus, when x is a rational, $\exp x = e^x$. If x is not a rational, then e^x has no meaning in terms of powers and roots. Therefore we make the following definition.

DEFINITION. *If x is irrational,* $e^x = \exp x$.

This definition, together with Theorem 7.2.7, implies that $e^x = \exp x$ for all real values of x. In view of Theorem 7.2.3 e^x obeys the usual rules of indices, whether x is rational or irrational. For example,

$$e^x \cdot e^y = e^{x+y}.$$

Functions defined in terms of the exponential function. If we require a function whose derivative is proportional to its size at any instant, then we may achieve this by defining

$$f(x) = \exp(kx).$$

Then
$$f'(x) = kf(x).$$

The hyperbolic functions, cosh and sinh, are defined in terms of the exponential function and are of practical importance. We are not concerned with their properties here, but we define them so that they may be used as examples.

DEFINITION.

$$\cosh x = \tfrac{1}{2}(\exp x + \exp(-x)).$$

DEFINITION.

$$\sinh x = \tfrac{1}{2}(\exp x - \exp(-x)).$$

Exercises 7.2

1. Use Theorems 7.2.2 and 7.2.4 to sketch the Cartesian graph of f when $f(x) = \exp x$.

2. Show from first principles that

(i) $x^2 e^x \to \infty$ as $x \to \infty$,

(ii) $x^2 e^x \to 0$ as $x \to -\infty$.

Sketch the Cartesian graph of f when $f(x) = x^2e^x$.
Where has f maxima or minima?

3. If $f(x) = e^{-x^2}$ and $g(x) = x^2e^{-x^2}$, discuss the behaviour of f and g as x tends to infinity and as x tends to minus infinity.
Sketch the Cartesian graphs of the functions f and g.

4. If $f(x) = \exp |x|$, is f (i) continuous, and (ii) differentiable at zero?

5. Use the definition of $\exp x$ to show that $e < 2.75$. Show also that $11/30 < 1/e < 3/8$.

6. Use the definition of a Maclaurin series to show that the Maclaurin series for $\exp x$ is

$$\sum_{n=0}^{\infty} \frac{x^n}{n!}.$$

7. If $f(x) = \sinh x$ and $g(x) = \cosh x$, show that f and g are differentiable everywhere and that $f'(x) = \cosh x$ and $g'(x) = \sinh x$.

8. Show that, if $f(x) = \sinh x$, then f is an increasing function and $f(x) \to \infty$ as $x \to \infty$, $f(x) \to -\infty$ as $x \to -\infty$.

9. Use Theorem 7.2.6 to evaluate

$$\lim_{x \to 0} \left[\frac{\sinh x}{x}\right].$$

10. Obtain the Maclaurin series for $\cosh x$ and for $\sinh x$.

7.3. Trigonometric functions

The usual definitions of trigonometric functions in terms of the ratios of sides of right-angled triangles are very evidently based upon geometric intutition. In particular, they are based upon the notion of angle, which is a very difficult concept to define.

Our intention in this section is to develop the theory of trigonometric functions quite independently of their association with triangles, or, indeed, with angles at all. The functions which we define will be shown to have the familiar properties of trigonometric functions; indeed, from an intuitive point of view, Theorem 6.5.2 will be sufficient to reassure us the functions we shall define are, indeed, the trigonometric functions with which the reader is familiar. However, from the formal point of view, we shall no longer assume the results of Theorems 5.2.5, 5.2.6, 5.4.3, 6.1.9, and 6.5.2 since the proofs of all these theorems depended upon the use of geometric intuition.

DEFINITION.

$$\sin x = \sum_{n=0}^{\infty} \frac{(-1)^n x^{2n+1}}{(2n+1)!}.$$

DEFINITION.

$$\cos x = \sum_{n=0}^{\infty} \frac{(-1)^n x^{2n}}{(2n)!}.$$

It is readily shown that the radius of convergence of both these series is infinity, so that $\sin x$ and $\cos x$ are defined for all real values of x. We note immediately that $\sin 0 = 0$ and that $\cos 0 = 1$.

We shall write $\cos^2 x$ for $(\cos x)^2$, $\sin^4 x$ for $(\sin x)^4$, and so on.

THEOREM 7.3.1. *If $f(x) = \sin x$ and $g(x) = \cos x$, then f and g are differentiable everywhere and*

$$f'(x) = \cos x \quad \textit{and} \quad g'(x) = -\sin x.$$

Proof. (i)

$$f(x) = \sum_{n=0}^{\infty} \frac{(-1)^n x^{2n+1}}{(2n+1)!}$$

with radius of convergence infinity. Therefore, by Theorem 7.1.4, f is differentiable everywhere and

$$f'(x) = \sum_{n=0}^{\infty} \frac{(-1)^n (2n+1) x^{2n}}{(2n+1)!} = \sum_{n=0}^{\infty} \frac{(-1)^n x^{2n}}{(2n)!} = \cos x.$$

(ii)

$$g(x) = \sum_{n=0}^{\infty} \frac{(-1)^n x^{2n}}{(2n)!}$$

with radius of convergence infinity. Therefore, by Theorem 7.1.4, g is differentiable everywhere and (putting $s = n-1$)

$$g'(x) = \sum_{n=0}^{\infty} \frac{(-1)^n (2n) x^{2n-1}}{(2n)!} = \sum_{s=0}^{\infty} \frac{(-1)^{s+1} x^{2s+1}}{(2s+1)!} = -\sin x.$$

THEOREM 7.3.2. (i) $\cos^2 x + \sin^2 x = 1$,

$$\text{(ii) } |\sin x| \leqslant 1, \, |\cos x| \leqslant 1$$

for all real values of x.

Proof. (i) Define the function f by

$$f(x) = \sin^2 x + \cos^2 x.$$

Then $$f'(x) = 2 \sin x \cdot \cos x + 2 \cos x \cdot -\sin x = 0.$$

Therefore, by Theorem 6.4.6, f is constant. Thus, for all x,

$$f(x) = \cos^2 0 + \sin^2 0 = 1.$$

Hence $$\cos^2 x + \sin^2 x = 1.$$

(ii) Is an immediate consequence of (i). □

THEOREM 7.3.3.

$$\sin(x+y) = \sin x \cdot \cos y + \cos x \cdot \sin y.$$

Proof. Let $a = x+y$. Define the function f by

$$f(x) = \sin x \cdot \cos(a-x) + \cos x \cdot \sin(a-x).$$

Then $$f'(x) = \cos x \cdot \cos(a-x) + \sin x \cdot \sin(a-x) - \sin x \cdot \sin(a-x) - \\ - \cos x \cdot \cos(a-x) = 0.$$

Therefore, by Theorem 6.4.6, f is constant. Thus, for all x,

$$f(x) = \sin 0 \cdot \cos a + \cos 0 \cdot \sin a = \sin a$$

Hence $\sin x \cdot \cos(a-x) + \cos x \cdot \sin(a-x) = \sin a$

or $\sin x \cdot \cos y + \cos x \cdot \sin y = \sin(x+y)$. □

THEOREM 7.3.4.

$$\cos(x+y) = \cos x \cdot \cos y - \sin x \cdot \sin y.$$

Proof. This theorem is proved in a similar manner to Theorem 7.3.3, by defining the function

$$f(x) = \cos x \cdot \cos(a-x) - \sin x \cdot \sin(a-x).$$

The proof is left to the reader. □

THEOREM 7.3.5.

$$\cos x = \cos(-x); \quad \sin x = -\sin(-x).$$

Proof. This theorem follows immediately from the definitions of $\sin x$ and $\cos x$. □

THEOREM 7.3.6(i). *There is a positive real number* u *such that* $\cos u = 0$.

(ii) *There is a least positive real number* v *such that* $\cos v = 0$.

Proof. (i) $\cos 0 = 1$

$$\cos 2 = 1 - \frac{2^2}{2!} + \frac{2^4}{4!} - \frac{2^6}{6!} + \frac{2^8}{8!} - \ldots - \frac{2^{4n-2}}{(4n-2)!} + \frac{2^4}{(4n)!} \cdots$$

$$= 1 - \frac{2^2}{2!} + \frac{2^4}{4!} - \frac{2^6}{6!}\left(1 - \frac{2^2}{56}\right) - \ldots - \frac{2^{4n-2}}{(4n-2)!}\left(1 - \frac{2^2}{(4n-1)\,4n}\right)$$

$$< 1 - \frac{2^2}{2!} + \frac{2^4}{4!}$$

$$= -\frac{1}{3}.$$

Therefore $\cos 2 < 0$.

Let $f(x) = \cos x$. f is continuous in view of Theorem 7.3.1. Therefore, by the intermediate value theorem, there is a point u in $(0, 2)$ such that $\cos u = 0$.

(ii) Let A be the set $\{u: \cos u = 0, u > 0\}$. We shall show that A must have a least member v.

A is bounded below by zero; so A has a greatest lower bound which we shall call v. Let $f(x) = \cos x$. Since f is continuous, $\cos v = \lim_{x \to u} \cos x$; since v is the *greatest* lower bound of A, there is a point u, either coinciding with v or as close as we please to v, such that $\cos u = 0$. Therefore $\lim \cos x = 0$ and so $\cos v = 0$. Therefore v belongs to A and is the least member of A. □

DEFINITION. *If v is the least positive number such that* $\cos v = 0$ *(v exists by Theorem* 7.3.6(ii)), *then we define* $\pi = 2v$. *Thus* $\cos \pi/2 = 0$, *but* $\cos x > 0$ *whenever* $0 \leqslant x < \pi/2$.

THEOREM 7.3.7. $\sin \pi/2 = 1$. $\cos \pi = -1$. $\sin \pi = 0$. $\cos 2\pi = 1$ $\sin 2\pi = 0$.

Proof. If $f(x) = \sin x$, then $f'(x) = \cos x$. Therefore f' is positive in $(0, \pi/2)$. Thus f is an increasing function in $[0, \pi/2]$ and, since $\sin 0 = 0$ $\sin \pi/2 > 0$. But

$$\sin^2 \pi/2 + \cos^2 \pi/2 = 1 \quad \text{and} \quad \cos \pi/2 = 0.$$

Therefore $\sin \pi/2 = 1$.

By Theorem 7.3.4,

$$\cos \pi = \cos \pi/2 \cdot \cos \pi/2 - \sin \pi/2 \cdot \sin \pi/2 = -1.$$

By Theorem 7.3.3,

$$\sin \pi = \sin \pi/2 \cdot \cos \pi/2 + \cos \pi/2 \cdot \sin \pi/2 = 0.$$

By Theorem 7.3.4,

$$\cos 2\pi = \cos \pi \cdot \cos \pi - \sin \pi \cdot \sin \pi = 1.$$

By Theorem 7.3.3,

$$\sin 2\pi = \sin \pi \cdot \cos \pi + \cos \pi \cdot \sin \pi = 0. \ \square$$

THEOREM 7.3.8. $\sin x$ *and* $\cos x$ *are periodic, with period* 2π; *in other words* $\sin(x+2\pi) = \sin x$ *and* $\cos(x+2\pi) = \cos x$ *for all real values of* x.

Proof. By Theorems 7.3.3 and 7.3.4,

$$\sin(x+2\pi) = \sin x \cdot \cos 2\pi + \cos x \cdot \sin 2\pi = \sin x,$$
$$\cos(x+2\pi) = \cos x \cdot \cos 2\pi - \sin x \cdot \sin 2\pi = \cos x. \ \square$$

DEFINITION. *If* $\cos x \neq 0$, *then* $\tan x = \sin x/\cos x$.

DEFINITION. *If* $\cos x \neq 0$, *then* $\sec x = 1/\cos x$.

DEFINITION. *If* $\sin x \neq 0$, *then* $\cot x = \cos x/\sin x$.

DEFINITION. *If* $\sin x \neq 0$, *then* $\operatorname{cosec} x = 1/\sin x$.

Other elementary identities involving trigonometric functions may be obtained in the usual way, by using these definitions and Theorems 7.3.1–7.3.8.

Inverse trigonometric functions. If $f(x) = \sin x$ and if the the domain of f is defined to be the interval $[-\pi/2, \pi/2]$, then f is a one-to-one mapping between the intervals $[-\pi/2, \pi/2]$ and $[-1, 1]$. Therefore there is an inverse function f^{-1} whose domain is $[-1, 1]$ and whose range is $[-\pi/2, \pi/2]$. We write

$$f^{-1}(x) = \sin^{-1}(x).$$

If x belongs to the interval $[-1, 1]$, then there are an infinity of real numbers y such that $\sin y = x$, but our definition of $\sin^{-1} x$ picks out a particular one of them, that one which lies between $-\pi/2$ and $\pi/2$.

So, too, if $g(x) = \cos x$ and if the domain of g is defined to be the interval $[0, \pi]$, then g is a one-to-one mapping between the intervals $[0, \pi]$ and $[-1, 1]$. Therefore there is an inverse function g^{-1} whose domain is $[-1, 1]$ and whose range is $[0, \pi]$. We write

$$g^{-1}(x) = \cos^{-1}(x).$$

If x belongs to the interval $[-1, 1]$, then there are an infinity of real numbers y such that $\cos y = x$, but our definition of $\cos^{-1} x$ picks out a particular one of them, that one which lies between 0 and π.

Finally, if $F(x) = \tan x$ and if the domain of F is defined to be the interval $(-\pi/2, \pi/2)$, then F is a one-to-one mapping between the interval $(-\pi/2, \pi/2)$ and the set of all real numbers. Therefore there is an inverse function F^{-1} whose domain is the set of real numbers and whose range is $(-\pi/2, \pi/2)$. We write

$$F^{-1}(x) = \tan^{-1}(x).$$

If x is any real number, then there are an infinity of real numbers y such that $\tan y = x$, but our definition of $\tan^{-1}(x)$ picks out a particular one of them—that one which lies between $-\pi/2$ and $\pi/2$.

THEOREM 7.3.9. (i) *If* $f(x) = \sin^{-1} x$, *then* f *is differentiable in the interval* $(-1, 1)$ *and*

$$f'(x) = \frac{1}{\sqrt{(1-x^2)}}.$$

(ii) *If* $f(x) = \cos^{-1} x$, *then* f *is differentiable in the interval* $(-1, 1)$ *and*

$$f'(x) = \frac{-1}{\sqrt{(1-x^2)}}$$

(iii) *If* $f(x) = \tan^{-1} x$, *then* f *is differentiable everywhere and*

$$f'(x) = \frac{1}{1+x^2}.$$

Proof. (i) Let $g(x) = \sin x$. Then $f = g^{-1}$. Let $y = \sin^{-1} x$. Then, by Theorem 6.1.4,

$$f'(x) = \frac{1}{g'(y)} = \frac{1}{\cos y} > 0, \quad \text{since } -\frac{\pi}{2} < y < \frac{\pi}{2}.$$

Therefore

$$f'(x) = \frac{1}{\sqrt{(1-\sin^2 y)}} = \frac{1}{\sqrt{(1-x^2)}}.$$

(ii) Let $g(x) = \cos x$. Then $f = g^{-1}$. Let $y = \cos^{-1} x$. Then, by Theorem 6.1.4,

$$f'(x) = \frac{1}{g'(y)} = \frac{1}{-\sin y} < 0, \quad \text{since } 0 < y < \pi.$$

Therefore

$$f'(x) = \frac{-1}{\sqrt{(1-\cos^2 y)}} = \frac{-1}{\sqrt{(1-x^2)}}.$$

(iii) Let $g(x) = \tan x$. Then $f = g^{-1}$. Let $y = \tan^{-1} x$. Then, by Theorem 6.1.4,

$$f'(x) = \frac{1}{g'(y)} = \frac{1}{\sec^2 y} = \frac{1}{1+\tan^2 y} = \frac{1}{1+x^2}. \quad \square$$

The main importance of inverse trigonometric functions is that they enable us to find anti-derivatives for certain functions (and therefore, as we shall see in the following chapter, they enable us to find the definite integrals of these functions). For example, the anti-derivative of

$$\frac{1}{\sqrt{(a^2-x^2)}} \quad \text{is} \quad \sin^{-1}\frac{x}{a}$$

and of

$$\frac{1}{a^2+x^2} \quad \text{is} \quad \frac{1}{a}\tan^{-1}\frac{x}{a}.$$

Exercises 7.3

1. Use Theorems 7.3.1–7.3.8 to deduce the following:

(i) $\sin 2x = 2 \sin x \cdot \cos x$.

(ii) $\sin (x-y) = \sin x \cdot \cos y - \cos x \cdot \sin y$.

(iii) $\cos (x-y) = \cos x \cdot \cos y + \sin x \cdot \sin y$.

(iv) $\cos (\pi - x) = -\cos x$.

(v) $\cos (2\pi - x) = \cos x$.

(vi) $\sin (\pi/2 + x) = \cos x$.

(vii) $\sin x + \sin y = 2 \sin \{(x+y)/2\} \cdot \cos \{(x-y)/2\}$.

(viii) $\sin 3x = 3 \sin x - 4 \sin^3 x$.

(ix) $\sin \pi/3 = \sqrt{3}/2$. $\cos \pi/3 = \frac{1}{2}$.

(x) The set of zeros of $\cos x$ is the set

$$\{u: u = (n+\tfrac{1}{2})\pi, n \text{ is an integer}\}.$$

(xi) The set of zeros of $\sin x$ is the set

$$\{u: u = n\pi, n \text{ is an integer}\}.$$

(xii) If $f(x) = \tan x$, then $f'(x) = \sec^2 x$.

(xiii) If $f(x) = \operatorname{cosec} x$, then $f'(x) = -\operatorname{cosec} x \cdot \cot x$.

2. If $f(x) = \tan x$,

(i) show that f is an increasing function in the interval $(-\pi/2, \pi/2)$,

(ii) show that $\tan (x+\pi) = \tan x$, whenever $\tan x$ is defined.
Sketch the Cartesian graph of f.

3. Use L'Hopital's rule to obtain the following limits:

(i) $\lim_{x \to 0} \left[\dfrac{\sin x}{x} \right]$.

(ii) $\lim_{x \to 0} \left[\dfrac{\tan x}{x} \right]$.

(iii) $\lim_{x \to 0} \left[\dfrac{\cos x - 1}{x^2} \right]$.

If

$$f(x) = \frac{\sin x}{x} \quad \text{when} \quad x \neq 0,$$

sketch the Cartesian graph of f.
What is the nature of the discontinuity at zero?

4. Use Theorem 4.9.2 to obtain the first four terms in the Maclaurin series for $e^x \cos x$.

5. In defining $\sin^{-1} x$ we stated that, if $f(x) = \sin x$, then f is a one-to-one mapping between the intervals $[-\pi/2, \pi/2]$ and $[-1, 1]$.
Prove this statement.

6. If $-1 \leqslant x \leqslant 1$, show that

$$\cos^{-1} x + \sin^{-1} x = \frac{\pi}{2}.$$

Explain the connection between this result and Theorem 7.3.9 (i) and (ii).

7.4. Logarithmic functions

If $f(x) = \exp x$, then f is defined for all real x and is strictly increasing; also $f(x) \to \infty$ as $x \to \infty$ and $f(x) \to 0$ as $x \to -\infty$. Therefore f is a one-to-one mapping between the sets $(-\infty, \infty)$ and $(0, \infty)$. Hence there is an inverse function mapping $(0, \infty)$ onto $(-\infty, \infty)$.

DEFINITION. *If* $f(x) = \exp x$, *then we write* $f^{-1}(x) = \log x$.

Hence $\log x$ is defined for all positive numbers x. $\log(\exp x) = x$ and $\exp(\log x) = x$. $\log 1 = 0$ and $\log e = 1$ (Fig. 7.1).

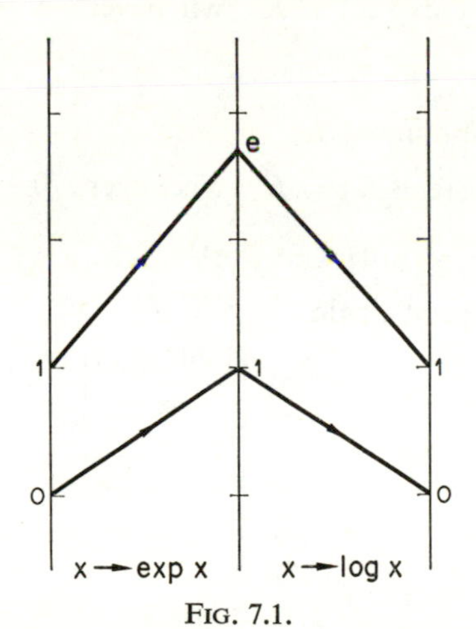

FIG. 7.1.

THEOREM 7.4.1. *If* $f(x) = \log x$, *then* f *is differentiable in the interval* $(0, \infty)$ *and* $f'(x) = 1/x$.

Proof. Let $g(x) = \exp x$. Then $f = g^{-1}$. Therefore, by Theorem 6.1.4,

$$f'(x) = \frac{1}{g'(\log x)} = \frac{1}{\exp(\log x)} = \frac{1}{x}. \square$$

THEOREM 7.4.2. (i) *If* $f(x) = \log x$, *then* f *is strictly increasing.* $\log x > 0$ *when* $x > 1$. $\log x < 0$ *when* $0 < x < 1$.

(ii) $\log x \to \infty$ *as* $x \to \infty$.

(iii) $\log x \to -\infty$ *as* $x \to 0+0$.

(iv) $\lim_{x \to 0} \left[\frac{\log(1+x)}{x}\right] = 1.$

Proof. (i) If $f(x) = \log x$, then f is defined when $x > 0$. $f'(x) = 1/x$; thus f' is positive. Therefore f is strictly increasing. The rest is an immediate consequence, since $\log 1 = 0$.

(ii) Given a real number K,

$$\log x > \log(\exp K) = K \quad \text{whenever} \quad x > \exp K.$$

Therefore $\log x \to \infty$ as $x \to \infty$.

(iii) Given a real number K,

$$\log x < \log(\exp K) = K \quad \text{whenever} \quad 0 < x < \exp K.$$

Therefore $\log x \to -\infty$ as $x \to 0+0$.

(iv) Using L'Hopital's rule,

$$\lim_{x \to 0} \left[\frac{\log(1+x)}{x}\right] = \lim_{x \to 0} \left[\frac{\frac{1}{1+x}}{1}\right] = 1.$$

THEOREM 7.4.3. *If x and y are positive real numbers, m and n are natural numbers and a is a rational, then*

(i) $\log 1/x = -\log x$.

(ii) $\log (xy) = \log x + \log y$.

(iii) $\log (x^{m/n}) = m/n \log x$.

(iv) $\log (x^a) = a \cdot \log x$.

Proof. (i) $\exp (\log 1/x + \log x) = \exp (\log 1/x) \cdot \exp (\log x) = (1/x) \cdot x = 1$.

Therefore $$\log (\exp(\log 1/x + \log x)) = \log 1 = 0.$$

Therefore $$\log 1/x + \log x = 0.$$

Hence $$\log 1/x = -\log x.$$

(ii) $$\exp (\log x + \log y) = \exp (\log x) \cdot \exp (\log y) = xy.$$

Therefore $$\log (\exp(\log x + \log y)) = \log (xy).$$

Hence $$\log x + \log y = \log (xy).$$

(iii) By repeated application of (ii),

$$\log (x^{p/q}) = p \cdot \log (x^{1/q}) \quad \text{and} \quad \log x = q \cdot \log (x^{1/q}).$$

Therefore $$\log (x^{p/q}) = p/q \cdot \log x.$$

(iv) This is an immediate consequence of (iii), (i), and the fact that $\log 1 = 0$.

x^a. When a is rational, Theorem 7.4.3(iv) gives

$$\log (x^a) = a \cdot \log x.$$

Thus $$x^a = \exp (\log(x^a)) = \exp (a \cdot \log x).$$

If a is not rational, then x^a has no meaning in terms of powers and roots. Therefore we make the following definition.

DEFINITION. *If a is irrational and if $x > 0$, then $x^a = \exp(a \cdot \log x)$.*

Thus, for all real values of a, $x^a = \exp(a \cdot \log x)$, either by definition or as a consequence of Theorem 7.4.3.

THEOREM 7.4.4. *If a, b and x are real numbers and if p and q are natural numbers and if $x > 0$, then*

(i) $x^{a+b} = x^a \cdot x^b$.

(ii) $(x^a)^{p/q} = x^{ap/q}$.

(iii) *If $f(x) = x^a$, f is differentiable in $(0, \infty)$ and*

$$f'(x) = ax^{a-1}.$$

Proof. (i)
$$\begin{aligned} x^{a+b} &= \exp((a+b) \cdot \log x) \\ &= \exp(a \cdot \log x + b \cdot \log x) \\ &= \exp(a \cdot \log x) \cdot \exp(b \cdot \log x) \\ &= x^a \cdot x^b. \end{aligned}$$

(ii)
$$\begin{aligned} (x^a)^{p/q} &= (\exp(a \cdot \log x))^{p/q} \\ &= \exp(pa/q \cdot \log x) \\ &= x^{pa/q}. \end{aligned}$$

(iii) If a is rational, then this result is a summary of Theorems 6.1.5–6.1.8. If a is irrational,

$$f(x) = x^a = \exp(a \cdot \log x).$$

Therefore
$$\begin{aligned} f'(x) &= \frac{a}{x} \cdot \exp(a \cdot \log x) \\ &= \frac{a}{x} \cdot x^a \\ &= ax^{a-1}. \quad \square \end{aligned}$$

DEFINITION. *When x and a are positive, $\log_a x$ is the power to which a has to be raised to equal x, i.e. $\log_a x$ is defined by $a^{\log_a x} = x$.*

THEOREM 7.4.5. (i) $\log_a x = \log x/\log a$.

(ii) *If* $f(x) = \log_a x$, *then* f *is differentiable in* $(0, \infty)$ *and*

$$f'(x) = \frac{1}{x \cdot \log a}.$$

(iii) $\log_a (xy) = \log_a x + \log_a y$.

(iv) *If* r *is a real number, then*

$$\log_a (x^r) = r \cdot \log_a x.$$

(v) $\log_a b = \dfrac{1}{\log_b a}$.

Proof. (i) $a^{\log_a x} = x \quad \text{or} \quad \exp(\log_a x \cdot \log a) = x$.

Thus $$\log(\exp(\log_a x \cdot \log a)) = \log x.$$

Therefore $$\log_a x \cdot \log a = \log x.$$

Hence $$\log_a x = \frac{\log x}{\log a}.$$

(ii) $f(x) = \log_a x = \dfrac{\log x}{\log a}$.

Therefore $$f'(x) = \frac{1}{x \cdot \log a}.$$

(iii), (iv), and (v) are left as exercises. □

COROLLARY. $\log_e x = \log x$.

Proof. This is an immediate consequence of Theorem 7.4.5(i), since $\log e = 1$. □

The log series. In order to evaluate $\log x$ for particular values of x we look for an expansion as a power series. It is clear that there can be no Maclaurin expansion for $\log x$, since log is not even defined at zero. We obtain instead the Maclaurin series for $\log(1+x)$.

THEOREM 7.4.6. *If* $-1 < x \leqslant 1$,

$$\log(1+x) = \sum_{n=1}^{\infty} \frac{(-1)^{n+1}x^n}{n}.$$

Proof. Let $f(x) = \log(1+x)$. Then

$$f'(x) = \frac{1}{1+x} \quad \text{and} \quad f^{(n)}(x) = \frac{(-1)^{n+1}(n-1)!}{(1+x)^n}.$$

Thus $$f^{(n)}(0) = (-1)^{n+1}(n-1)!$$

Taylor's theorem now gives

$$\log(1+x) = x - \frac{x^2}{2} + \frac{x^3}{3} - \ldots + \frac{(-1)^n x^{n-1}}{n-1} + R_n.$$

We now have to show that, when $-1 < x \leqslant 1$, $R_n \to 0$ as $n \to \infty$. When $x = 0$ this is trivial. When $0 < x \leqslant 1$ we use the Lagrange form of remainder.

$$R_n = \frac{(-1)^{n+1}(n-1)!\,x^n}{n!\,(1+\theta x)^n} \quad \text{where } \theta \text{ depends upon } n \text{ and } 0 < \theta < 1.$$

Thus $$|R_n| = \frac{x^n}{n(1+\theta x)^n} < \frac{1}{n}.$$

Therefore $R_n \to 0$ as $n \to \infty$.

When $-1 < x < 0$ we are no longer able to use the Lagrange form of remainder, since $1+\theta x$ is no longer greater than 1. Therefore we use the Cauchy form

$$R_n = \frac{(-1)^{n+1}(n-1)!\,(1-\phi)^{n-1}x^n}{(n-1)!\,(1+\phi x)^n} \quad \text{where } \phi \text{ depends upon } n \text{ and } 0 < \phi < 1.$$

Thus
$$|R_n| = \frac{(1-\phi)^{n-1}|x|^n}{(1+\phi x)^n}$$
$$= \frac{|x|^n}{1-\phi}\left(\frac{1-\phi}{1+\phi x}\right)^n.$$

Now $|\phi x| < \phi$. Therefore $\dfrac{1-\phi}{1+\phi x} < 1$. Thus $|R_n| < \dfrac{|x|^n}{1-\phi}$.

Therefore $R_n \to 0$ as $n \to \infty$.

Hence, if $-1 < x \leqslant 1$,
$$\log(1+x) = \sum_{n=1}^{\infty} \frac{(-1)^{n+1}x^n}{n}. \quad \square$$

As the reader will have noticed in the proof of the previous theorem, showing that the remainder term of a Maclaurin series does tend to zero is often quite difficult. It is almost always easier to show that the Maclaurin series is convergent, e.g. it is easily shown that the radius of convergence of the Maclaurin series in the previous theorem is 1 and hence that the series is convergent when $-1 < x \leqslant 1$. However, as we have previously remarked on p. 183, proving that a Maclaurin series is convergent is *not* the same as proving that its sum to infinity gives the original function; we still need to show that R_n tends to zero and not to some other finite limit. Thus Theorem 7.4.6 is necessary and may not be replaced by a simple proof of convergence. The reader is referred to exercises 7.4, question 5 for an example of a function whose Maclaurin series converges for all values of x but which, nevertheless, converges to the wrong sum!

Evaluation of logarithms. The Maclaurin series may be used in principle to evaluate $\log x$ when $0 < x \leqslant 2$. Its convergence is, however, extremely slow, so that in practice we use a modified form. We have
$$\log(1+x) = x - \frac{x^2}{2} + \frac{x^3}{3} - \frac{x^4}{4} + \dots$$

and
$$\log(1-x) = -x-\frac{x^2}{2}-\frac{x^3}{3}-\frac{x^4}{4}-\dots$$

when $|x| < 1$. We now write

$$\frac{1+x}{1-x} = \frac{m}{n} \quad \text{or} \quad (m+n)x = m-n$$

so that

$$\log\frac{m}{n} = 2\left[\frac{m-n}{m+n}+\frac{(m-n)^3}{3(m+n)^3}+\frac{(m-n)^5}{5(m+n)^5}+\frac{(m-n)^7}{7(m+n)^7}+\dots\right].$$

This result holds for all natural numbers m and n.

For example, if we put $m = 2$ and $n = 1$, we obtain

$$\log 2 = 2\left[\frac{1}{3}+\frac{1}{3.3^3}+\frac{1}{5.3^5}+\frac{1}{7.3^7}+\frac{1}{9.3^9}+\dots\right].$$

If we wish to evaluate log 20, we may obtain a series for log 20 by putting $m = 20$ and $n = 1$, but the convergence of this series is too slow to be of practical use. We therefore obtain log 20 indirectly. For example,

$$\log 20 = 4\log 2+\log \tfrac{5}{4},$$

and both of the logarithms on the right-hand side may be conveniently evaluated by means of the series.

To evaluate other logarithms, say logarithms to base 10, we first evaluate log 10 and then use Theorem 7.4.5(i).

Logarithms and the evaluation of limits

THEOREM 7.4.7. *l is a positive real number. $f(x) \to l$ as $x \to a$ if and only if $\log f(x) \to \log l$ as $x \to a$.*

Proof. Suppose, first that $f(x) \to l$ as $x \to a$. Let $g(x) = \log x$. We may define or re lefine $f(a) = l$ without affecting the value of $\lim_{x\to a} f(x)$.

Then f is continuous at a. Also g is continuous at l. Therefore, by Theorem 5.4.6, $g \circ f$ is continuous at a; in other words,

$$\log(f(x)) \to \log l \quad \text{as} \quad x \to a.$$

Suppose now that $\log(f(x)) \to \log l$ as $x \to a$. Let $g(x) = \exp x$. We may again define or redefine $f(a) = l$ without affecting the value of $\lim_{x \to a} \log(f(x))$. Then, writing $h(x) = \log(f(x))$, h is continuous at a. Also g is continuous at $\log l$. Therefore, by Theorem 5.4.6, $g \circ h$ is continuous at a; in other words

$$\exp(\log(f(x))) \to \exp(\log l) \quad \text{as} \quad x \to a$$

or

$$f(x) \to l \quad \text{as} \quad x \to a. \ \square$$

THEOREM 7.4.8. *l is a positive real number. $f(x) \to l$ as $x \to \infty$ if and only if* $\log(f(x)) \to \log l$ *as $x \to \infty$.*

Proof. If $x > 0$ we redefine $f(-x) = f(x)$. This does not affect the value of

$$\lim_{x \to \infty} f(x).$$

Then

$$\lim_{x \to \infty} f(x) = \lim_{x \to 0} f\left(\frac{1}{x}\right)$$

and

$$\lim_{x \to \infty} \log(f(x)) = \lim_{x \to 0} \log\left(f\left(\frac{1}{x}\right)\right).$$

The result now follows from Theorem 7.4.7. $\square$

COROLLARY. *$\{a_n\}$ is a sequence of positive real numbers. $a_n \to l$ as $n \to \infty$ if and only if* $\log a_n \to \log l$ *as $n \to \infty$.*

Proof. The function f is defined by

$$f(x) = a_{[x]} \quad \text{if} \quad x \geqslant 1$$

The result now follows from Theorem 7.4.8. $\square$

These theorems are useful in the evaluation of limits of functions or sequences involving powers or factorials. By way of example we obtain the important result

$$\lim_{n\to\infty}\left(1+\frac{x}{n}\right)^n = e^x.$$

We have

$$\lim_{n\to\infty} \log\left[\left(1+\frac{x}{n}\right)^n\right]$$

$$= \lim_{n\to\infty} n \log\left(1+\frac{x}{n}\right)$$

$$= \lim_{n\to\infty} x\,\frac{\log(1+x/n)}{x/n}$$

$$= x$$

Theorem 7.4.2(iv). Therefore, by the corollary to Theorem 7.4.8,

$$\lim\left(1+\frac{x}{n}\right)^n = e^x.$$

Logarithmic differentiation. The logarithmic function is is often used to evaluate the derivatives of functions for which the variable appears in the index. The method is based upon the following theorem.

THEOREM 7.4.9. *f is a differentiable function and $f(x) > 0$ everywhere in the interval (a, b). If g is defined throughout the interval (a, b) by*

$$g(x) = \log(f(x))$$

then g is differentiable in (a, b) and

$$f'(x) = f(x)g'(x).$$

Proof. Let $h(x) = \log x$. Then $g = h \circ f$. Therefore, by Theorem 6.1.3,

$$g'(x) = h'(f(x))\cdot f'(x).$$

But $$h'(f(x)) = \frac{1}{f(x)}.$$

Therefore $$g'(x) = \frac{f'(x)}{f(x)}. \square$$

The way in which this theorem is used to obtain derivatives is illustrated by the following example.

If $f(x) = x^x$, define $g(x) = \log(x^x) = x \cdot \log x$.

Therefore $g'(x) = \log x + 1$.

Therefore $f'(x) = f(x) \cdot g'(x) = (\log x + 1)x^x$.

Exercises 7.4

1. Evaluate the following limits:

(i) $\lim_{x \to 0} \left[\frac{\log(1+2x)}{x}\right]$. (ii) $\lim_{x \to 1} \frac{\log_2 x}{\tan(3x-3)}$

2. Prove Theorem 7.4.5 (iii), (iv) and (v).

3. Use the first four terms of the series

$$\log 2 = 2\left[\frac{1}{3} + \frac{1}{3.3^3} + \frac{1}{5.3^5} + \frac{1}{7.3^7} + \frac{1}{9.3^9} + \ldots\right]$$

to evaluate log 2 correct to four decimal places. Prove that your answer is correct to four decimal places, i.e. show that the sum to infinity of the remaining terms is not large enough to affect the fourth decimal place.

4. Using the result of question 2 as necessary, evaluate log 3, log 6, and log 167 to three decimal places. Show that your results are all correct to three decimal places.

5. Evaluate $\log_{10} 2$ to four decimal places.

6. The function f is defined by

$$\begin{cases} f(x) = \exp\left(-\frac{1}{x^2}\right) & \text{if} \quad x \neq 0, \\ f(0) = 0. \end{cases}$$

Show that f is differentiable as many times as we please at zero and that

$$f^{(n)}(0) = 0.$$

Thus we obtain the Maclaurin series $0+0+0+0+0+0+\ldots$ which converges but which certainly does not give $f(x)$.

7. Use Theorem 3.6.3 and the identity

$$\frac{\log(2^n)}{(2^n)^a} = \frac{n\cdot\log 2}{(2^a)^n}$$

to evaluate

$$\lim_{x\to\infty}\left[\frac{\log x}{x^a}\right]$$

when $a > 0$.

8. Find the derivative of f in each of the following cases:

(i) $f(x) = x^{\sin x}$. (iii) $f(x) = (\sin x)^x$.

(ii) $f(x) = 2^x$. (iv) $f(x) = x^{x^x}$.

In each case state also the domain of definition of f.

9. Show that

$$\frac{1}{3\cdot\log 3}+\frac{1}{4\cdot\log 4} > \frac{1}{4\cdot\log 2}$$

$$\frac{1}{5\cdot\log 5}+\frac{1}{6\cdot\log 6}+\frac{1}{7\cdot\log 7}+\frac{1}{8\cdot\log 8} > \frac{1}{6\cdot\log 2}$$

$$\frac{1}{9\cdot\log 9}+\ldots+\frac{1}{16\cdot\log 16} > \frac{1}{8\cdot\log 2}$$

Use the behaviour of the harmonic series to prove that

$$\sum_{n=2}^{\infty}\frac{1}{n\cdot\log n}$$

is divergent.

10. If $f(x) = \log x$, use the results of Theorems 7.4.1 and 7.4.2 to sketch the Cartesian graph of f.

11. If $f(x) = (\log x)/x$, use the results of Theorems 7.4.1 and 7.4.2 and question 6 to sketch the Cartesian graph of f.

7.5. Infinite products

The English mathematician John Wallis obtained the following formula for π:

$$\frac{\pi}{2} = \frac{2}{1}\cdot\frac{2}{3}\cdot\frac{4}{3}\cdot\frac{4}{5}\cdot\frac{6}{5}\cdot\frac{6}{7}\cdot\frac{8}{7}\cdot\frac{8}{9}\cdot\ldots.$$

The right-hand side of this equation is an infinite product. Infinite products are analogous to infinite series, successive terms being multiplied instead of added. In this section we shall discuss infinite products briefly (Wallis's formula will be proved in the following chapter on p. 264).

DEFINITION. *If b_n is a sequence, then*

$$b_1 \cdot b_2 \cdot b_3 \cdot b_4 \cdot b_5 \ldots.$$

is called an infinite product. *We shall write the product Πb_n.*

DEFINITION. *The product*

$$b_1 \cdot b_2 \cdot b_3 \ldots b_n$$

is called the nth *partial product of the infinite product b_n. It is written compactly $\prod_{r=1}^{n} b_r$ and is often denoted by P_n.*

DEFINITION. *Πb_n is an infinite product whose* nth *partial product will be called P_n. If $\{P_n\}$ is a convergent sequence and if* $\lim P_n = l \neq 0$, *then we say that the infinite product Πb_n is convergent and equal to l. We write $\Pi b_n = l$.*

Why are we not prepared to say that an infinite product is convergent if the limit of its sequence of partial products is zero? In order that an infinite series should converge we demand that ultimately one partial sum does not differ much from another. This leads us to the conclusion that the limit of the nth term is zero. In order that an infinite product should converge, we demand similarly that ultimately one partial product does not differ much from another. This leads us to the conclusion that the limit of the nth term is one (this result is proved formally in the next theorem). If one of the terms of the infinite product is zero or if, for sufficiently large n, $|b_n| \leqslant K < 1$, then the limit of the partial products will be zero. In order to preserve the analogy with infinite series we regard this as a case of divergence.

THEOREM 7.5.1. *If the infinite product $\Pi(1+a_n)$ is convergent, then* $\lim a_n = 0$.

Proof. Let P_n be the nth partial product. Since $\Pi(1+a_n)$ is convergent, $\{P_n\}$ is also convergent. Let $\lim P_n = l$. Then $\lim P_{n-1} = l$. Therefore, by the corollary to Theorem 3.3.3, $\lim (1+a_n) = 1$. Hence $\lim a_n = 0$. □

We note that this theorem for infinite products is analogous to Theorem 4.2.4 for infinite series. The converse of this theorem, like the converse of Theorem 4.2.4, is not true.

The following theorem gives a criterion for the convergence of an infinite product and emphasizes the connection between infinite products and series.

THEOREM 7.5.2. *Suppose Σa_n^2 is convergent and $a_n \neq -1$ for all n. Then Σa_n is convergent if and only if $\Pi(1+a_n)$ is convergent.*

Proof. Since Σa_n^2 is convergent, $|a_n| < 1$ for all sufficiently large n. Then

$$\log(1+a_n) = a_n - \frac{a_n^2}{2} + \frac{a_n^3}{3} - \frac{a_n^4}{4} + \frac{a_n^5}{5} - \frac{a_n^6}{6} + \frac{a_n^7}{7} - \ldots$$

$$= a_n - a_n^2\left(\frac{1}{2} - \frac{a_n}{3}\right) - a_n^4\left(\frac{1}{4} - \frac{a_n}{5}\right) - a_n^6\left(\frac{1}{6} - \frac{a_n}{7}\right) - \ldots$$

$$= a_n - \frac{a_n^2}{2} + a_n^3\left(\frac{1}{3} - \frac{a_n}{4}\right) + a_n^5\left(\frac{1}{5} - \frac{a_n}{6}\right) + a_7^n\left(\frac{1}{7} - \frac{a_n}{8}\right) + \ldots.$$

Therefore $\qquad 0 < a_n - \log(1+a_n) < \frac{1}{2}a_n^2.$

But Σa_n^2 is convergent. Therefore, by the comparison test, $\Sigma a_n - \log(1+a_n)$ is convergent. Therefore, by Theorem 4.2.5, Σa_n is convergent if and only if $\Sigma \log(1+a_n)$ is convergent. But

$$\log\left(\prod_{r=1}^{n} (1+a_r)\right) = \sum_{r=1}^{n} \log(1+a_r).$$

Therefore, by the corollary to Theorem 7.4.8, $\Sigma \log(1+a_n)$ is convergent if and only if $\Pi(1+a_n)$ is convergent. Hence Σa_n is convergent if and only if $\Pi(1+a_n)$ is convergent. □

Exercises 7.5

1. Show that the infinite product

$$\frac{3}{2}\cdot\frac{3}{4}\cdot\frac{15}{8}\cdot\frac{9}{16}\cdot\frac{63}{32}\cdot\frac{33}{64}\cdot\frac{255}{128}\cdot\frac{129}{256}\cdot\frac{1023}{512}\cdots$$

is divergent.

2. Show that the infinite product

$$\frac{3}{2}\cdot\frac{2}{3}\cdot\frac{5}{4}\cdot\frac{4}{5}\cdot\frac{7}{6}\cdot\frac{6}{7}\cdot\frac{9}{8}\cdot\frac{8}{9}\cdots$$

is convergent and find its value. Use the result of Theorem 4.3.1 to deduce that

$$1-\tfrac{1}{2}+\tfrac{1}{3}-\tfrac{1}{4}+\tfrac{1}{5}-\tfrac{1}{6}+\tfrac{1}{7}\cdots$$

is convergent.

3. Prove that

$$\Pi\left(1+\frac{1}{n^{2/3}}\right)$$

is divergent.

7.6. The binomial theorem

The binomial theorem is the name used to describe two quite different results which are formally similar but which are of very different degrees of complexity. In this section we prove these two results in order to contrast their levels of sophistication; the first is elementary, the second is certainly not.

THEOREM 7.6.1 (the binomial theorem for positive integral index). *If a and b are real numbers and n is a natural number, then*

$$(a+b)^n = {}_nC_0a^n+{}_nC_1a^{n-1}b+\ldots+{}_nC_ra^{n-r}b^r+\ldots+{}_nC_nb^n,$$

where ${}_nC_r = n!/r!(n-r)!$ *is the number of ways of choosing r objects from n.*

Proof. $(a+b)^n = (a+b)(a+b)\ldots(a+b)$. $a^{n-r}b^r$ will appear in the product exactly the same number of times as the number of ways of choosing r b's from n brackets, i.e. ${}_nC_r$. Thus

$$(a+b)^n = {}_nC_0a^n + \ldots + {}_nC_ra^{n-r}b^r + \ldots + {}_nC_nb^n. \square$$

THEOREM 7.6.2 (the binomial theorem when the index *is not* a positive integer). *If a is a real number and if* $|x| < 1$, *then*

$$(1+x)^a = 1 + \sum_{r=1}^{\infty} \frac{a(a-1)\ldots(a-r+1)}{r!} x^r.$$

Proof. Let $f(x) = (1+x)^a$. Then $f'(x) = a(1+x)^{a-1}$,

$$f^{(n)}(x) = a(a-1)(a-2)\ldots(a-n+1)(1+x)^{a-n}.$$

Taylor's theorem with the Cauchy form of remainder now gives

$$(1+x)^a = 1 + ax + \frac{a(a-1)}{2!}x^2 + \ldots + \frac{a(a-1)\ldots(a-n+2)}{(n-1)!}x^{n-1} + R_n$$

where $$R_n = \frac{a(a-1)\ldots(a-n+1)}{(n-1)!}(1+\phi x)^{a-n}(1-\phi)^{n-1}x^n,$$

$$|R_n| = \left|\frac{a(a-1)\ldots(a-n+1)}{(n-1)!}\right|(1+\phi x)^{a-1}\left(\frac{1-\phi}{1+\phi x}\right)^{n-1}|x|^n$$

$$< \left|\frac{a(a-1)\ldots(a-n+1)}{(n-1)!}\right| 2^{a-1}|x|^n.$$

We leave the reader to verify, as an exercise, that, if $|x| < 1$,

$$\lim_{n\to\infty}\left[\frac{a(a-1)\ldots(a-n+1)}{(n-1)!}x^n\right] = 0.$$

It follows that $\lim R_n = 0$, which completes the proof. $\square$

Exercises 7.6

1. If
$$P_n = \frac{a(a-1)\dots(a-n+1)}{(n-1)!}x^n, \text{ what is } \frac{P_{n+1}}{P_n}\text{ ?}$$

Deduce that
$$\lim P_n = 0 \quad \text{when} \quad |x| < 1.$$

2. Show that, when $|x| > 1$, the binomial series
$$1+\sum_{r=1}^{\infty} \frac{a(a-1)\dots(a-r+1)}{r!}x^r$$
is divergent by use of D'Alembert's ratio test.

Use the same test to show that the series is convergent when $|x| < 1$. Explain carefully why this does not provide an alternative method of dealing with Theorem 7.6.2.

Exercises for Chapter 7

1. If a function is defined for all real x by $f(x) = e^{-x}\cos x$, find $f(0), f'(x), f'(0)$. Show that (a) e^{-x} is positive for all real x, (b) $e^{-x} < 1$ if $x > 0$.
Using $-1 \leqslant \cos x \leqslant 1$, deduce inequalities for $f(x)$.
Sketch a graph of $f(x)$. [T.C.]

2. Sketch the graph of $f(x)=1/\log x$. Show that $g(x)=1/(x \log x)$ is monotonic decreasing for $x > 1$. Hence or otherwise show that the series
$$\sum_{n=2}^{\infty} \frac{(-1)^n}{n \log n}$$
is convergent.

Prove that the series $\sum_{n=1}^{\infty} \frac{1}{n}$ is divergent and that the series $\sum_{n=2}^{\infty} \frac{1}{n \log n}$ is divergent. [T.C.]

3. State what is meant by the expression "$f(x)$ is differentiable at the point $x = a$".
Prove that, if $f(x)$ is differentiable at $x = a$, then it is continuous at $x = a$.

If
$$f(x) = \begin{cases} \dfrac{e^x-1}{x}, & x \neq 0, \\ 1, & x = 0, \end{cases}$$
find (a) $f'(1)$, (b) $f'(0)$. (Do not assume continuity of $f'(x)$ at $x = 0$.) [T.C.]

4. State Taylor's theorem, with Lagrange's form of remainder, for a function $f(x)$ which is n times differentiable for $a \leqslant x \leqslant b$. By the application of this theorem, derive a power series for $\log(1+x)$ and prove that this expansion of the function is valid in the range $[0, 1]$.

State what difficulty arises in establishing, by this means, the validity of the same expansion in the interval $(-1, 0]$.

Obtain an expansion for $\log(1+x^2)$ as an infinite series and state its range of validity. [T.C.]

5. Write down the Maclaurin expansion for e^x and prove *from the series* that

$$\lim_{x \to \pm\infty} \frac{P(x)}{e^{x^2}} = 0,$$

where $P(x)$ is any polynomial in x.

Sketch the graphs of e^{kx^2} in the cases $k > 0$, $k = 0$, $k < 0$, and hence show that the equation

$$e^{kx^2} = 2 \cos x$$

has a finite even number of roots when $k > 0$ and an infinite number of roots otherwise.

Show that, if n is a positive integer, then there is just one positive value of k for which the equation has $4n$ solutions but there are infinitely many positive values of k for which the equation has $4n+2$ solutions. [B.Sc.]

6. (i) Explain what is meant by the statement "Σa_r is convergent".

What conclusions can you draw from applying the ratio test for absolute convergence to the binomial series

$$1+nx+\frac{n(n-1)}{2!}x^2+\frac{n(n-1)(n-2)}{3!}x^3+\ldots?$$

(ii) Investigate, for all real values of x, the convergence or divergence of

(a) $\cosh x+\cosh 2x+\cosh 3x+\ldots+\cosh nx+\ldots$.

(b) $1+2^2x+3^2x^2+\ldots+n^2x^{n-1}\ldots$.

(iii) Show that $1+(1/2^2)+(1/3^2)+\ldots$ is convergent and hence or otherwise show that $\Sigma(n/(n^3+1))$ also converges. [T.C.]

7. The series $\sum_{n=1}^{\infty} a_n$ of positive terms is convergent. Show that there exists an integer N such that $a_n^{1/n} \leqslant 1$ for $n \geqslant N$.

If the series $\sum_{n=1}^{\infty} a_n x^n$ is convergent for $x = r > 0$, show that it is also convergent for $|x| < r$. Deduce further that the series $\sum_{n=1}^{\infty} na_n x^{n-1}$ converges for $|x| < r$.

Find the sum of the series

$$\sum_{n=1}^{\infty} n^2x^n \quad \text{for} \quad |x| < 1.$$

$\left(\text{You may assume that } \frac{d}{dx}\left(\sum_{n=1}^{\infty} a_nx^n\right) = \sum_{n=1}^{\infty} na_nx^{n-1}\right)$ [B.Sc.]

8. Show that

$$(n+1)\log n > n\log(n+1)$$

for every positive integer $n \geqslant 3$ and deduce that the sequence $\{\sqrt[n]{n}\}$ is decreasing for $n \geqslant 3$.

Deduce that if $\sum_{n=1}^{\infty} a_n$ is a convergent series of positive terms, then the series $\sum_{n=1}^{\infty} (\sqrt[n]{n})a_n$ is also convergent.

Does the series $\sum_{n=1}^{\infty} (\sqrt[n]{n})/n$ converge or diverge? Give reasons. [B.Sc.]

9. $f(x)$ is defined, for $|x| \leqslant 1$, as the sum of the series $\sum_{n=1}^{\infty} x^n/n^2$. Prove that, in the range of definition:

(i) $$f(x)+f(-x) = \tfrac{1}{2}f(x^2)$$

(ii) $f(x)$ is a strictly increasing function.

(iii) $f(x) \geqslant x$.

(You should quote in full any general theorems on power series that you use.) [B.Sc.]

10. Show that the power series

$$1+\sum_{n=1}^{\infty} \frac{a(a-1)\ldots(a-n+1)}{n!}x^n,$$

where a is a fixed real number, converges for $|x| < 1$. Denoting by $f(x)$ the sum of this series for $|x| < 1$, show that $(1+x)f'(x) = af(x)$.

(Quote any theorems on power series that you use to prove this.)

Deduce that $f(x) = (1+x)^a$ for $|x| < 1$.

Find

$$\lim_{t\to 0} \frac{(1+t/3)^{1/2}-(1+t/2)^{1/3}}{(1-t/3)^{-1/2}-(1-t/2)^{-1/3}}.$$ [B.Sc].

CHAPTER 8

THE RIEMANN INTEGRAL

8.1. Introduction

In this chapter we turn to the problem of area and its measurement. This opening section will be concerned with the area of sets of points in the plane. Usually, the set whose area we wish to find will be a region, i.e. a set whose boundary is a simple closed curve (a continuous, closed curve which does not intersect itself). It is, in fact, always possible to ascribe to any region a definite area, although we shall not be able to prove this. Our approach in this section will depend very much upon the use of geometric intuition; in other words, this section is to be regarded as an informal introduction and we shall not begin the formal development of the theory until section 8.2.

We may conveniently choose as our unit of area any region of given shape and size which tessellates the plane; thus we might choose triangles, squares, parallelograms, regular hexagons, and so on. It is, of course, customary to use the square as the unit of area.

The measure of area which we shall develop, with the square as a unit, will satisfy the following, intuitively necessary, principles:

(i) *The area of any region is a positive real number.*

(ii) *If A and B are two regions such that $A \cap B = \varnothing$, then area of $A \cup B$ = area of A+area of B.*

(iii) *If A and B are two regions such that $A \subset B$, then area of $A \leqslant$ area of B.*

(iii) is in fact a consequence of (ii).

Area of a rectangle. If we have a rectangle of length m units and width n units, where m and n are natural numbers, then a straightforward count convinces us that we have mn squares. Thus the area of the rectangle is mn.

If we have a rectangle of length a units and width b units, where a and b are rationals, then we may cover the rectangle by squares and recognizable fractions of squares and find that the area of the rectangle is ab.

If now a rectangle, which we shall call R, has length x units and width y units, where at least one of x and y is irrational, then it is not possible to recognize intuitively the areas of the parts of squares necessary to cover the rectangle. In this case we use a limiting process based upon the principles of areal measure stated above.

If a, b, c, and d are rationals and if

$$a < x < b, \quad c < x < d,$$

then a rectangle with sides a and c may be inscribed in R and R may be inscribed in a rectangle with sides b and d. Assuming that it is meaningful to ascribe a definite area to the rectangle R the third principle stated above then gives

$$ac \leqslant \text{area of } R \leqslant bd.$$

Suppose that we now take four sequences of rationals $\{a_n\}$, $\{b_n\}$, $\{c_n\}$, and $\{d_n\}$, so that, for all n,

$$a_n < x < b_n, \quad c_n < y < d_n$$

and so that

$$\lim a_n = \lim b_n = x, \quad \lim c_n = \lim d_n = y.$$

By Theorem 3.3.3, $\lim (a_n c_n) = xy, \quad \lim (b_n d_n) = xy.$

But $$a_n c_n \leqslant \text{area of } R \leqslant b_n d_n.$$

Therefore, by the corollary to Theorem 3.3.6, area of $R = xy$.

In this way we are able to assign a measure to the area of any rectangle.

Areas of polygons. Once we are able to assign a measure of area to any rectangle, we may, in the usual way, deduce first measures for the areas of parallelograms, then of triangles, and, finally, of arbitrary polygons, by application of the principles set out above.

Areas of sets of points. We now give a definition of the area of a set A of points in the plane. This definition is applicable to arbitrary sets of points in the plane, defined in any manner, not just to regions.

Let S be the set of all polygons containing A and let T be the set of all polygons contained in A. Let k be the least upper bound of the areas of members of T and let K be the greatest lower bound of the areas of members of S. Then $k \leqslant K$. If $k = K$, then we say that the set A is measurable and that the area of A is k.

We note that not all sets of points are measurable. All regions of the plane are measurable, although we shall not be able to prove this.

Area of a circle. We shall now use this definition of the area of a set of points to find the area of a circle.

In the previous chapter we defined π in terms of the function $f(x) = \cos x$. In this informal section we shall have in mind the more familiar definition of π as the ratio of the lengths of the circumference and diameter of a circle. This definition, like the proof for the area of a circle which we are about to give, depends heavily upon the use of geometric intuition; indeed, the concept of length of arc, which figures prominently in both definition and proof, will not even be defined in this book.

THEOREM 8.1.1. *The area of a circle of radius r is πr^2.*

Proof. Inscribe in the circle a regular polygon of n sides, as in Fig. 8.1. Let the length of one side of the polygon be l and let its area be A_n. Then

$$A_n = n \cdot \tfrac{1}{2} l \sqrt{(r^2 - \tfrac{1}{4} l^2)}.$$

We assume that $nl \to 2\pi r$ as $n \to \infty$. Also $l \to 0$ as $n \to \infty$. Thus

$$A_n \to 2\pi r \cdot \tfrac{1}{2} r = \pi r^2 \quad \text{as} \quad n \to \infty.$$

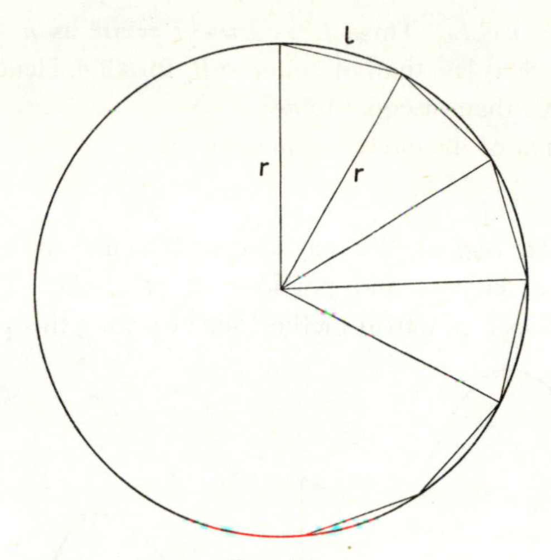

FIG. 8.1.

But the area of the circle is greater than or equal to A_n for all n. Hence the area of the circle is greater than or equal to πr^2.

Now inscribe the circle in a regular polygon of n sides, as in Fig. 8.2. Let the length of one side of the polygon be L and let its area be

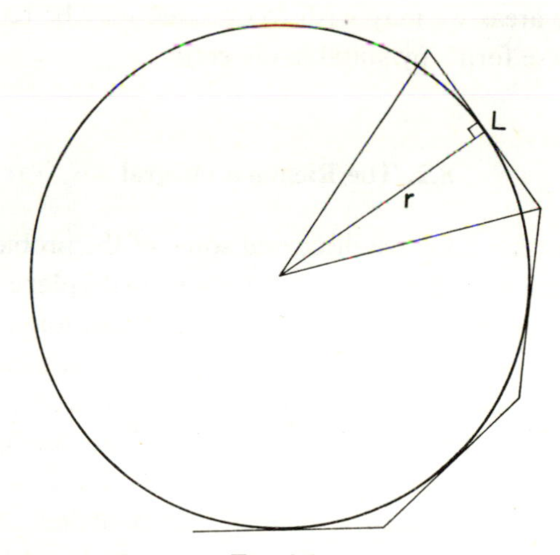

FIG. 8.2.

B_n. Then $B_n = n \cdot \frac{1}{2}Lr$. Thus $B_n \to 2\pi r \cdot \frac{1}{2}r = \pi r^2$ as $n \to \infty$. But the area of the circle is less than or equal to B_n for all n. Hence the area of the circle is less than or equal to πr^2.

Thus the area of the circle is equal to πr^2. □

Areas of other regions. We may also wish to find the areas of other shapes, such as ellipses and portions of parabolae. The Riemann integral provides a powerful method for attacking this problem. It is

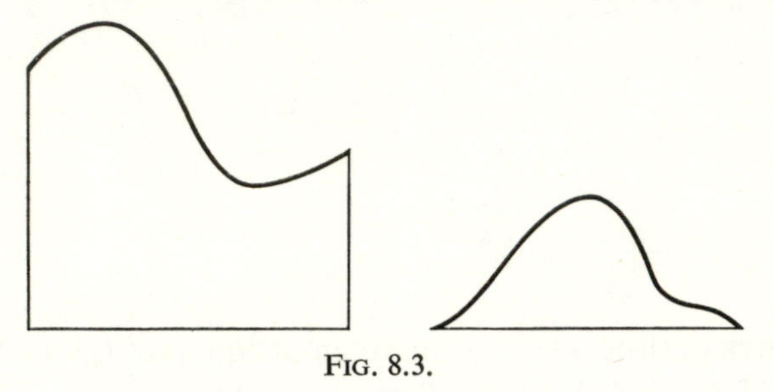

FIG. 8.3.

concerned with the evaluation of areas of regions consisting of one of the forms shown in Fig. 8.3.

Most of the areas we may wish to evaluate may be reduced to one or other of these forms by suitable dissection.

8.2. The Riemann integral

In the previous section we discussed some of the problems involved in assigning measures of area to sets of points in the plane, and suggested means of solving these problems. The discussion was informal, since may of the concepts we employed remained undefined, and some of the results we made use of were not formally proved. In this section we formally define a Riemann integral and begin to develop the formal theory of areal measure.

Suppose that f is a function defined and bounded in the interval $[a, b]$. (It is natural to think of f as being continuous; however, this restriction will not be imposed.) The Riemann integral of f will be the area of that region of the Cartesian graph of f indicated in Fig. 8.4. In order to define this area formally we inscribe and circumscribe polygons and then use the principles enunciated in section 8.1. In fact the polygons we use are sets of rectangles, as shown in Fig. 8.5. The

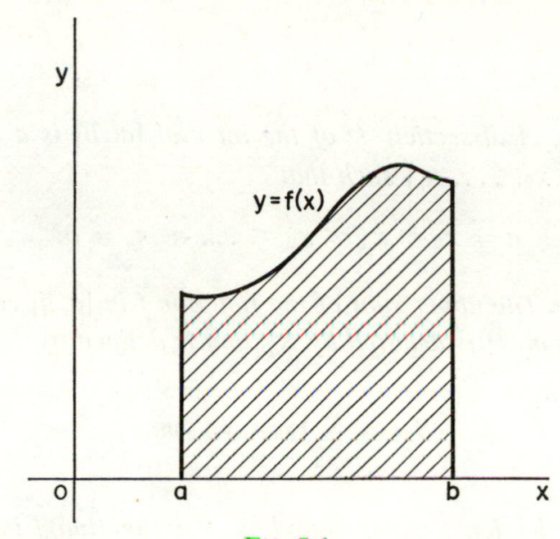

FIG. 8.4.

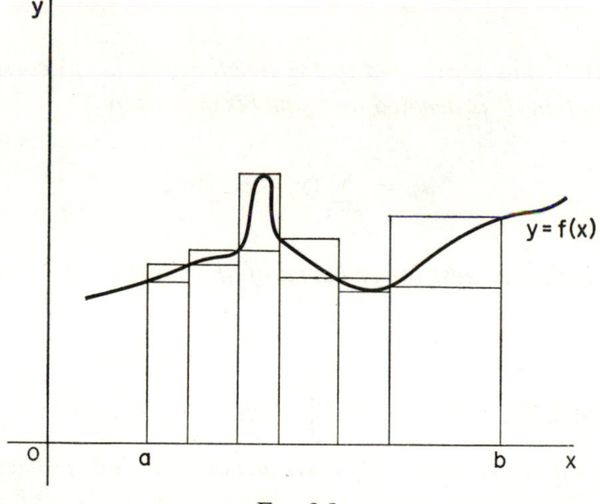

FIG. 8.5.

following definitions and theorems are the results of stating in formal language the intuitive idea that the area we require is greater than or equal to the area of any set of rectangles inscribed in the region and is less than or equal to the area of any set of rectangles circumscribing the region. This intuitive idea should be borne in mind as the reader proceeds.

DEFINITION. *A dissection* $\mathcal{D}$ *of the interval* $[a, b]$ *is a set of points* $\mathcal{D} = \{x_0, x_1, x_2, \ldots, x_n\}$ *such that*

$$a = x_0 \leqslant x_1 \leqslant x_2 \leqslant \ldots \leqslant x_n = b.$$

DEFINITION. *The upper sum of the function* f *in* $[a, b]$ *corresponding to the dissection* $\mathcal{D}$ *is denoted by* $S_{\mathcal{D}}$ *and is defined by*

$$S_{\mathcal{D}} = \sum_{r=1}^{n} (x_r - x_{r-1})m,$$

where M_r *is the least upper bound of the function* f *in the interval* $[x_{r-1}, x_r]$.

DEFINITION. *The lower sum of the function* f *in* $[a, b]$ *corresponding to the dissection* $\mathcal{D}$ *is denoted by* $s_{\mathcal{D}}$ *and is defined by*

$$s_{\mathcal{D}} = \sum_{r=1}^{n} (x_r - x_{r-1})m_r$$

where m_r *is the greatest lower bound of the function* f *in the interval* $[x_{r-1}, x_r]$.

DEFINITION. *The upper integral* $\overline{\int_a^b} f\,dx$ *is the greatest lower bound of the set of all upper sums* $S_{\mathcal{D}}$ *corresponding to all possible dissections* $\mathcal{D}$.

DEFINITION. *The lower integral* $\underline{\int_a^b} f\,dx$ *is the least upper bound of the set of all lower sums* $s_{\mathcal{D}}$ *corresponding to all possible dissections* $\mathcal{D}$.

DEFINITION. *If* $\overline{\int_a^b} f\,dx = \underline{\int_a^b} f\,dx$, *then* f *is said to be Riemann integrable in the interval* $[a, b]$. *In this case* $\int_a^b f\,dx$ *is called the Riemann integral of* f *in the interval* $[a, b]$ *and is defined by* $\int_a^b f\,dx = \overline{\int_a^b} f\,dx = \underline{\int_a^b} f\,dx$.

This sequence of definitions serves to define the area of the region indicated in Fig. 8.4. We now investigate some of the properties of the objects which we have defined.

THEOREM 8.2.1. *If* $\mathcal{D}$ *is any dissection of* $[a, b]$ *and if* $S_{\mathcal{D}}$ *and* $s_{\mathcal{D}}$ *are the upper and lower sums of the function* f *corresponding to* $\mathcal{D}$, *then*

$$\text{(i) } S_{\mathcal{D}} \geqslant s_{\mathcal{D}}, \quad \text{(ii) } S_{\mathcal{D}} \geqslant \overline{\int_a^b} f\,dx, \quad \text{(iii) } s_{\mathcal{D}} \leqslant \underline{\int_a^b} f\,dx.$$

Proof. (i) $S_{\mathcal{D}} - s_{\mathcal{D}} = \sum_{r=1}^{n} (M_r - m_r)(x_r - x_{r-1}) \geqslant 0$.

(ii) and (iii) are immediate consequences of the definitions of upper and lower integrals. □

THEOREM 8.2.2. *If* $\mathcal{D}$ *is a dissection of* $[a, b]$ *and if* $\mathcal{D}'$ *is a dissection obtained from* $\mathcal{D}$ *by adding one point to* $\mathcal{D}$, *then*

$$\text{(i) } S_{\mathcal{D}} \geqslant S_{\mathcal{D}'}, \quad \text{(ii) } s_{\mathcal{D}} \leqslant s_{\mathcal{D}'}.$$

Proof. (i) Let y be the point added and let $x_{r-1} \leqslant y \leqslant x_r$. Then $S_{\mathcal{D}'}$ is the same as $S_{\mathcal{D}}$ except that $M_r(x_r - x_{r-1})$ is replaced by $P_r(y - x_{r-1}) + Q_r(x_r - y)$, where P_r is the least upper bound of f in the interval $[x_{r-1}, y]$ and Q_r is the least upper bound of f in the interval $[y, x_r]$. Now

$$P_r \leqslant M_r \quad \text{and} \quad Q_r \leqslant M_r.$$

Therefore
$$P_r(y - x_{r-1}) + Q_r(x_r - y) \leqslant M_r(y - x_{r-1} + x_r - y)$$
$$= M_r(x_r - x_{r-1}).$$

Thus $S_{\mathcal{D}} \geqslant S_{\mathcal{D}'}$.
(ii) is proved similarly. □

THEOREM 8.2.3. *If $\mathcal{D}$ and $\mathcal{D}^*$ are two dissections of* $[a, b]$, then

$$\text{(i) } S_{\mathcal{D} \cup \mathcal{D}^*} \leqslant S_{\mathcal{D}}, \quad \text{(ii) } s_{\mathcal{D} \cup \mathcal{D}^*} \geqslant s_{\mathcal{D}}.$$

Proof. This theorem follows from Theorem 8.2.2 simply by adding the points of $\mathcal{D}^*$ one by one. □

THEOREM 8.2.4. *If $\mathcal{D}$ and $\mathcal{D}^*$ are two dissections of* $[a, b]$, *then* $S_{\mathcal{D}} \geqslant s_{\mathcal{D}^*}$, *i.e. the upper sums all exceed the lower sums.*

Proof. By Theorems 8.2.1 and 8.2.3,

$$S_{\mathcal{D}} \geqslant S_{\mathcal{D} \cup \mathcal{D}^*} \geqslant s_{\mathcal{D} \cup \mathcal{D}^*} \geqslant s_{\mathcal{D}^*}. \; \square$$

THEOREM 8.2.5. $\overline{\int_a^b} f\,dx \geqslant \underline{\int_a^b} f\,dx.$

Proof. Given $\varepsilon > 0$, we can find a dissection $\mathcal{D}$ with

$$S_{\mathcal{D}} - \overline{\int_a^b} f\,dx < \frac{\varepsilon}{2}$$

and we can find a dissection $\mathcal{D}^*$ with

$$\underline{\int_a^b} f\,dx - s_{\mathcal{D}^*} < \frac{\varepsilon}{2}$$

Thus $$S_{\mathcal{D}}-s_{\mathcal{D}*}-\left(\overline{\int_a^b} f\,dx-\underline{\int_a^b} f\,dx\right)<\varepsilon.$$

$$\text{Therefore } \overline{\int_a^b} f\,dx-\underline{\int_a^b} f\,dx>S_{\mathcal{D}}-s_{\mathcal{D}*}-\varepsilon>-\varepsilon.$$

Thus $$\overline{\int_a^b} f\,dx-\underline{\int_a^b} f\,dx,$$

which is a fixed quantity independent of the choice of ε, exceeds any negative number $-\varepsilon$. Thus

$$\overline{\int_a^b} f\,dx-\underline{\int_a^b} f\,dx\geqslant 0. \square$$

THEOREM 8.2.6. *If $a<b<c$,*

(i) $\overline{\int_a^b} f\,dx+\overline{\int_b^c} f\,dx=\overline{\int_a^c} f\,dx.$

(ii) $\underline{\int_a^b} f\,dx=\underline{\int_b^c} f\,dx=\underline{\int_a^c} f\,dx.$

If, in addition, f is Riemann integrable in $[a, c]$, then

(iii) $\int_a^b f\,dx+\int_b^c f\,dx=\int_a^c f\,dx.$

Proof. (i) The idea behind the proof of this theorem is simple but the proof is technically complicated to write out.

Given $\varepsilon>0$, there is a dissection $\mathcal{D}$ of $[a, b]$ such that

$$S_{\mathcal{D}}-\overline{\int_a^b} f\,dx<\frac{\varepsilon}{2}.$$

There is also a dissection $\mathcal{E}$ of $[b, c]$ such that

$$S_{\mathcal{E}}-\overline{\int_b^c} f\,dx<\frac{\varepsilon}{2}$$

Now $\mathcal{D}\cup\mathcal{E}$ is a dissection of $[a, c]$ and

$$S_{\mathcal{D}\cup\mathcal{E}} = S_{\mathcal{D}}+S_{\mathcal{E}}.$$

Therefore $\overline{\int_a^c} f\,dx \leqslant S_{\mathcal{D}\cup\mathcal{E}} = S_{\mathcal{D}}+S_{\mathcal{E}} \leqslant \overline{\int_a^b} f\,dx + \overline{\int_b^c} f\,dx+\varepsilon.$

But all the upper integrals are definite quantities independent of the choice of ε. Thus

$$\overline{\int_a^c} f\,dx \leqslant \overline{\int_a^b} f\,dx + \overline{\int_b^c} f\,dx.$$

Given $\varepsilon > 0$, there is a dissection $\mathcal{F}$ of $[a, c]$ such that

$$S_{\mathcal{F}} - \overline{\int_a^c} f\,dx < \varepsilon.$$

Let $\mathcal{F}'$ be the dissection obtained from $\mathcal{F}$ by adding the point b. By Theorem 8.2.2 $S_{\mathcal{F}'} \leqslant S_{\mathcal{F}}$, so that

$$S_{\mathcal{F}'} - \overline{\int_a^c} f\,dx < \varepsilon.$$

Let $\mathcal{D}'$ and $\mathcal{E}'$ be the dissections of $[a, b]$ and $[b, c]$ respectively induced by $\mathcal{F}'$ (so that $\mathcal{F}' = \mathcal{D}'\cup\mathcal{E}'$). Then

$$S_{\mathcal{D}'}+S_{\mathcal{E}'} = S_{\mathcal{F}'}.$$

Thus $\overline{\int_a^b} f\,dx + \overline{\int_b^c} f\,dx \leqslant S_{\mathcal{D}'}+S_{\mathcal{E}'} = S_{\mathcal{D}'\cup\mathcal{E}'} \leqslant \overline{\int_a^c} f\,dx+\varepsilon.$

But all the upper integrals are definite quantities independent of the choice of ε. Thus

$$\overline{\int_a^b} f\,dx + \overline{\int_b^c} f\,dx \leqslant \overline{\int_a^c} f\,dx.$$

Hence $$\overline{\int_a^b} f\,dx + \overline{\int_b^c} f\,dx = \overline{\int_a^c} f\,dx.$$

(ii) is proved similarly.

(iii) follows immediately from (i) in view of the definition of a Riemann integral. □

THEOREM 8.2.7. *If f and g are Riemann integrable functions in the interval $[a, b]$ and if the function F is defined by*

$$F(x) = f(x) + g(x)$$

for all x in $[a, b]$, then F is Riemann integrable in $[a, b]$ and

$$\int_a^b F\,dx = \int_a^b f\,dx + \int_a^b g\,dx.$$

Proof. Let $\mathcal{D} = \{x_0, x_1, x_2, \ldots, x_n\}$ be a dissection of $[a, b]$. We shall denote by $S_{\mathcal{D}}(f)$ the upper sum of the function f corresponding to the dissection $\mathcal{D}$, $M_r(f)$ the least upper bound f in the interval $[x_{r-1}, x_r]$, and so on. Then

$$M_r(F) \leqslant M_r(f) + M_r(g) \quad \text{for all } r.$$

Thus
$$S_{\mathcal{D}}(F) \leqslant S_{\mathcal{D}}(f) + S_{\mathcal{D}}(g).$$

So too
$$m_r(F) \geqslant m_r(f) + m_r(g),$$

giving
$$s_{\mathcal{D}}(F) \geqslant s_{\mathcal{D}}(f) + s_{\mathcal{D}}(g).$$

These inequalities hold for all possible dissections $\mathcal{D}$, so that

$$\underline{\int_a^b} f\,dx + \underline{\int_a^b} g\,dx \leqslant \underline{\int_a^b} F\,dx \leqslant \overline{\int_a^b} F\,dx \leqslant \overline{\int_a^b} f\,dx + \overline{\int_a^b} g\,dx.$$

But f and g are Riemann integral, giving

$$\underline{\int_a^b} f\,dx = \overline{\int_a^b} f\,dx = \int_a^b f\,dx$$

and
$$\underline{\int_a^b} g\,dx = \overline{\int_a^b} g\,dx = \int_a^b g\,dx.$$

Hence
$$\underline{\int_a^b} F\,dx = \overline{\int_a^b} F\,dx.$$

Thus F is Riemann integrable and

$$\int_a^b F\,dx = \int_a^b f\,dx + \int_a^b g\,dx. \quad \square$$

THEOREM 8.2.8. (i) *If $c > 0$, and if $g(x) = c \cdot f(x)$ for all x in $[a, b]$, then*

$$\overline{\int_a^b} g\,dx = c \cdot \overline{\int_a^b} f\,dx \quad \text{and} \quad \underline{\int_a^b} g\,dx = c \cdot \underline{\int_a^b} f\,dx.$$

(ii) *If $c < 0$ and if $g(x) = c \cdot f(x)$ for all x in $[a, b]$, then*

$$\overline{\int_a^b} g\,dx = c \cdot \underline{\int_a^b} f\,dx \quad \text{and} \quad \underline{\int_a^b} g\,dx = c \cdot \overline{\int_a^b} f\,dx.$$

(iii) *If f is Riemann integrable in $[a, b]$ and if $g(x) = c \cdot f(x)$ for all x in $[a, b]$, then g is Riemann integrable in $[a, b]$ and $\int_a^b g\,dx = c \cdot \int_a^b f\,dx$.*

The proof of this theorem is left to the reader.

THEOREM 8.2.9. (i) *If f is Riemann integrable in $[a, b]$ and if $f(x) \geqslant$ for all x in $[a, b]$, then $\int_a^b f\,dx \geqslant 0$.*

(ii) *If f is Riemann integrable in $[a, b]$, then so is $|f|$ and*

$$\int_a^b |f|\,dx \geqslant \left| \int_a^b f\,dx \right|.$$

Proof. (i) Given any dissection $\mathcal{D}$, $m_r \geqslant 0$ for all r. Thus $s_{\mathcal{D}} \geqslant 0$ and hence $\int_a^b f\,dx \geqslant 0$.

(ii) Firstly, $|f|$ may be shown to be Riemann integrable by proving that, given any dissection $\mathcal{D}$,

$$S_{\mathcal{D}}(|f|) - s_{\mathcal{D}}(|f|) \leqslant S_{\mathcal{D}}(f) - s_{\mathcal{D}}(f),$$

and by observing that, given $\varepsilon > 0$, a dissection $\mathcal{D}$ may be found such that $S_{\mathcal{D}}(f) - s_{\mathcal{D}}(f) < \varepsilon$. The inequality between the integrals may be proved by showing that, for any dissection $\mathcal{D}$,

$$|S_{\mathcal{D}}(f)| \leqslant S_{\mathcal{D}}(|f|).$$

The details are left to the reader. □

COROLLARY. *If f and g are Riemann integrable in $[a, b]$ and if $f(x) \geqslant g(x)$ for all x in $[a, b]$, then*

$$\int_a^b f\,dx \geqslant \int_a^b g\,dx.$$

Proof. Define F by $F(x) = f(x) - g(x)$ for all x in $]a, b]$. By Theorems 8.2.7 and 8.2.8,

$$\int_a^b F\,dx = \int_a^b f\,dx - \int_a^b g\,dx.$$

By Theorem 8.2.9,
$$\int_a^b F\,dx \geqslant 0.$$

Hence
$$\int_a^b f\,dx \geqslant \int_a^b g\,dx. \quad \square$$

Evaluation of the Riemann integral for specific functions. We illustrate the concept of the Riemann integral further by evaluating a particular Riemann integral.

Consider the integral $\int_0^h f\,dx$ when $f(x) = x^2$. It is customary to write this integral simply as $\int_0^h x^2\,dx$. We choose a dissection $\mathcal{D}$ of the form

$$\left\{0, \frac{h}{n}, \frac{2h}{n}, \frac{3h}{n}, \ldots, \frac{nh}{n} = h\right\}.$$

Then
$$S_{\mathcal{D}} = \sum_{r=1}^{n} \left(\frac{rh}{n}\right)^2 \frac{h}{n} = \frac{h^3}{n^3} \sum_{r=1}^{n} r^2.$$

We now use the result $\sum_{r=1}^{n} r^2 = \dfrac{n(n+1)(2n+1)}{6}$

and obtain
$$S_{\mathcal{D}} = \frac{h^3}{6} \cdot \left(1 + \frac{1}{n}\right)\left(2 + \frac{1}{n}\right).$$

Thus $$\overline{\int_0^h} x^2\,dx \leqslant \frac{h^3}{6}\cdot\left(1+\frac{1}{n}\right)\left(2+\frac{1}{n}\right)$$

for all values of n. But

$$\frac{h^3}{6}\cdot\left(1+\frac{1}{n}\right)\left(2+\frac{1}{n}\right) \to \frac{h^3}{3} \quad \text{as } n \to \infty.$$

Therefore, by Theorem 3.3.6,

$$\overline{\int_0^h} x^2\,dx \leqslant \frac{h^3}{3}.$$

So, too,

$$s_{\mathcal{D}} = \sum_{r=1}^{n}\left[\frac{(r-1)h}{n}\right]^2 \frac{h}{n} = \frac{h^3}{n^3}\sum_{r=1}^{n}(r-1)^2 = \frac{h^3}{n^3}\,\frac{(n-1)\,n(2n-1)}{6}$$
$$= \frac{h^3}{6}\cdot\left(1-\frac{1}{n}\right)\left(2-\frac{1}{n}\right).$$

Thus $$\underline{\int_0^h} x^2\,dx \geqslant \frac{h^3}{6}\cdot\left(1-\frac{1}{n}\right)\left(2-\frac{1}{n}\right)$$

for all values of n. Therefore, by Theorem 3.3.6,

$$\underline{\int_0^h} x^2\,dx \geqslant \frac{h^3}{3}.$$

So $$\frac{h^3}{3} \leqslant \underline{\int_0^h} x^2\,dx \leqslant \overline{\int_0^h} x^2\,dx \leqslant \frac{h^3}{3}.$$

Thus x^2 is Riemann integrable in the interval $[0, h]$ and

$$\int_0^h x^2\,dx = \frac{h^3}{3}.$$

Exercises 8.2

1. If $f(x) = 5$ evaluate upper and lower sums corresponding to any dissection of the interval [3, 6]. Hence show from the definition of the Riemann integral that f is Riemann integrable in the interval [3, 6] and that $\int_3^6 5\,dx = 15$.

2. Prove the following by mathematical induction:

(i) $\sum_{r=1}^{n} r = \frac{n(n+1)}{2}$. (ii) $\sum_{r=1}^{n} r^2 = \frac{n(n+1)(2n+1)}{6}$. (iii) $\sum_{r=1}^{n} r^3 = \frac{n^2(n+1)^2}{4}$.

3. $\int_2^7 x\,dx$ is the area of a certain trapezium and is $22\frac{1}{2}$. Show this by drawing a Cartesian graph of the function $f(x) = x$.

Now evaluate the integral formally using the definition of a Riemann integral and the result of question 2 (i).

4. Use the definition of the Riemann integral and the result of question 2 (iii) to evaluate

$$\int_0^h x^3\,dx.$$

Use Theorem 8.2.6 to deduce the value of

$$\int_2^5 x^3+2x\,dx.$$

5. Evaluate the following:

(i) $\int_{-3}^{4} |x|\,dx$. (ii) $\int_0^5 [x]\,dx$ (see exercises 5.2 question 2 (vi).)

(iii) $\int_{-5}^{5} f\,dx$, where f is defined by

$$\begin{cases} f(x) = x & \text{if } x < 0, \\ f(x) = 2 & \text{if } x \geqslant 0. \end{cases}$$

6. Under what circumstances is $\int_a^b f\,dx$ less than zero?

7. Prove Theorem 8.2.2 (ii).

8. Prove Theorem 8.2.8.

9. Complete the details of the proof of Theorem 8.2.9 (ii).

10. Use Theorems 8.2.7 and 8.2.8 to deduce, from the results of questions 3 and 4, the value of

$$\int_2^6 x(x-2)(x+2)\,dx.$$

8.3. Integrability of monotonic functions

Until now we have been concerned with developing the concept of a Riemann integral and examining particular functions for integrability. In this section and the one following we establish criteria for integrability.

A condition which is satisfied, in suitable intervals, by almost all commonly occurring functions, is that of being monotonic. It turns out that all monotonic functions are integrable.

THEOREM 8.3.1. *If f is monotonic in $[a, b]$, then f is Riemann integrable in $[a, b]$.*

Proof. Suppose that f is increasing. (The proof when f is decreasing is left to the reader.) f is bounded in $[a, b]$, since $f(a) \leqslant f(x) \leqslant f(b)$. Given $\varepsilon > 0$, choose n so that

$$\frac{f(b)-f(a)}{n} < \frac{\varepsilon}{b-a}.$$

Let $x_0 = a$ and, if $r > 0$, let x_r be the least upper bound of the set

$$\left\{x : f(x) \leqslant f(a) + \frac{r(f(b)-f(a))}{n}\right\},$$

as in Fig. 8.6. The reader is left to verify that

$$\mathcal{D} = \{x_0, x_1, x_2, x_3, \ldots, x_n\}$$

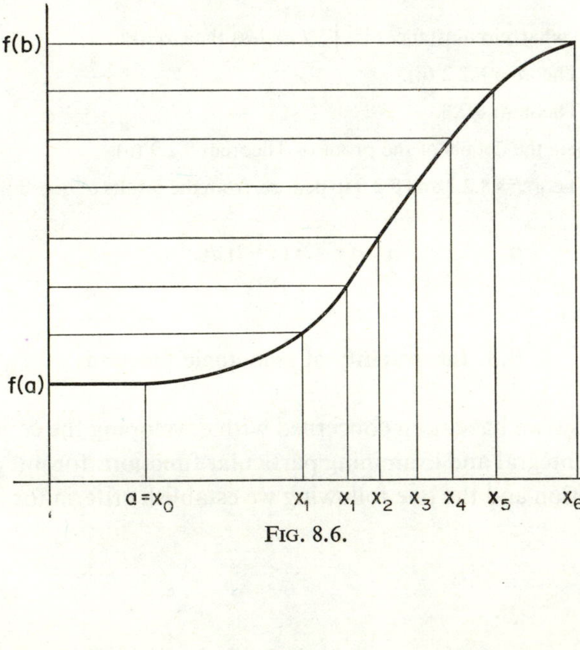

FIG. 8.6.

is a dissection of $[a, b]$. If x lies in $[x_{r-1}, x_r]$, then

$$f(a)+\frac{(r-1)\left(f(b)-f(a)\right)}{n} \leqslant f(x) \leqslant f(a)+\frac{r\left(f(b)-f(a)\right)}{n}.$$

Thus $$M_r-m_r \leqslant \frac{f(b)-f(a)}{n} < \frac{\varepsilon}{b-a}.$$

Now
$$\begin{aligned}\overline{\int_a^b} f\,dx-\underline{\int_a^b} f\,dx &\leqslant S_{\mathcal{D}}-s_{\mathcal{D}} \\ &\leqslant \sum_{r=1}^{n}(M_r-m_r)(x_r-x_{r-1}) \\ &< \frac{\varepsilon}{b-a}\sum_{r=1}^{n}(x_r-x_{r-1}) \\ &= \frac{\varepsilon}{b-a}(b-a) = \varepsilon.\end{aligned}$$

But ε is arbitrary, so that

$$\overline{\int_a^b} f\,dx-\underline{\int_a^b} f\,dx \leqslant 0.$$

Thus $$\overline{\int_a^b} f\,dx = \underline{\int_a^b} f\,dx \text{ and } f \text{ is integrable.} \quad \square$$

Theorems 8.3.1 and 8.2.6 together tell us that most commonly occurring functions are integrable since most are either monotonic or may be divided by one or more points into monotonic pieces. For example, the constant function and the functions f defined by

$$f(x) = x^3,\ f(x) = x,\ f(x) = \exp x,\ f(x) = \log x$$

are all monotonic, the functions defined by

$$f(x) = x^2,\ f(x) = x^4,\ f(x) = \exp x^2$$

can all be divided at $x = 0$ into two monotonic pieces. In any closed interval $[a, b]$ the functions f given by

$$f(x) = \sin x,\ f(x) = \cos x$$

may both be divided into a finite number of monotonic pieces. Thus all these functions are integrable in any closed interval.

Exercises 8.3

1. Complete the proof of Theorem 8.3.1.

2. Give further examples of functions which are known to be integrable in any closed interval by the use of Theorems 8.3.1 and 8.2.6.

3. If f is defined by

$$\begin{cases} f(x) = x \sin 1/x & \text{if } x \neq 0, \\ f(0) = 0, \end{cases}$$

explain why Theorems 8.3.1 and 8.2.6 do not immediately show that f is integrable in the interval [0, 1].

By considering

$$\int_{\varepsilon}^{1} f\,dx,$$

show that f is integrable in [0, 1].

8.4. Continuous functions and the Riemann integral

In the last section we saw that all monotonic functions are integrable. In this section we shall see that all continuous functions are integrable. Further, although continuity is not a necessary condition for integrability (some discontinuous functions are integrable), it is a condition of central importance, since it enables integrals to be simply evaluated.

THEOREM 8.4.1 (the first mean value theorem for integrals). *If f is a function, bounded in $[a, b]$, whose least upper bound is M and whose greatest lower bound is m, then*

$$(b-a)m \leqslant \underline{\int_a^b} f\,dx \leqslant \overline{\int_a^b} f\,dx \leqslant (b-a)M.$$

Proof. If $\mathcal{D}$ is the dissection $\{a, b\}$, then

$$S_{\mathcal{D}} = (b-a)M, \quad s_{\mathcal{D}} = (b-a)m.$$

The result follows, by Theorem 8.2.1 (and Theorem 8.2.5). □

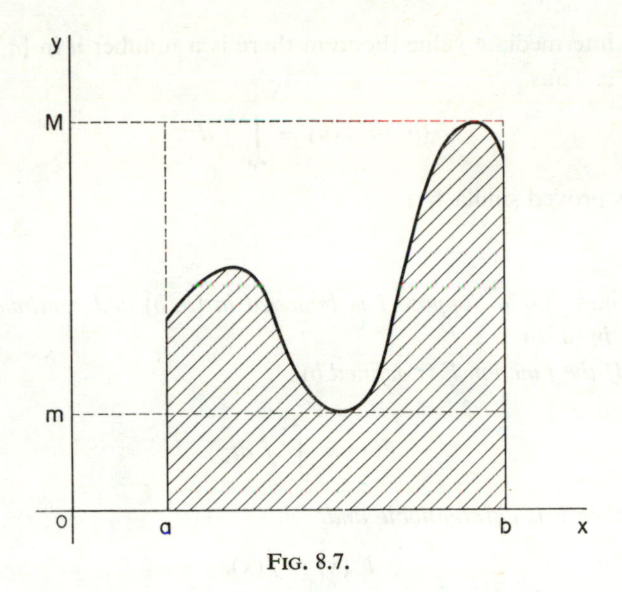

FIG. 8.7.

The result is intuitively obvious as Fig, 8.7 suggests.

THEOREM 8.4.2 (the second mean value theorem for integrals). *If the function f is continuous in the closed interval* $[a, b]$, *then we can find numbers u and v in* $[a, b]$ *such that*

$$\text{(i) } (b-a)f(u) = \overline{\int_a^b} f\,dx, \quad \text{(ii) } (b-a)f(v) = \underline{\int_a^b} f\,dx.$$

Proof. (i) Since f is continuous in $[a, b]$, f is bounded (by Theorem 5.7.2), so that

$$\overline{\int_a^b} f\,dx \quad \text{and} \quad \underline{\int_a^b} f\,dx$$

are both defined. We deduce from Theorem 8.4.1 that there is a real number c, satisfying $m \leqslant c \leqslant M$, such that

$$(b-a)\,c = \overline{\int_a^b} f\,dx.$$

By the intermediate value theorem there is a number u in $[a, b]$ with $f(u) = c$. Thus

$$(b-a)f(u) = \overline{\int_a^b} f\,dx.$$

(ii) is proved similarly. □

THEOREM 8.4.3. *Suppose f is bounded in $[a, b]$ and continuous at a point x in (a, b).*

(i) *If the function F is defined by*

$$F(x) = \overline{\int_a^x} f\,dt,$$

then F is differentiable and

$$F'(x) = f(x).$$

(ii) *If the function G is defined by*

$$G(x) = \underline{\int_a^x} f\,dt,$$

then G is differentiable and

$$G'(x) = f(x).$$

Now suppose f is continuous in $[a, b]$.

(iii) *Then f is Riemann integrable in $[a, b]$.*

Proof. (i) $$\frac{F(x+h)-F(x)}{h} = \frac{1}{h}\left[\overline{\int_a^{x+h}} f\,dt - \overline{\int_a^x} f\,dt\right]$$
$$= \frac{1}{h}\overline{\int_x^{x+h}} f\,dt$$
$$= \frac{h}{h}f(u)$$

for some number u in $[x, x+h]$.

As $h \to 0$, $u \to x$. Thus, since f is continuous at $x, f(u) \to f(x)$. Therefore

$$\lim_{h \to 0} \left[\frac{F(x+h)-F(x)}{h}\right] = f(x),$$

that is,

$$F'(x) = f(x).$$

(ii) follows similarly.

(iii) By the first two parts,

$$F'(x) = f(x) = G'(x).$$

Therefore

$$F'(x) - G'(x) = 0.$$

Hence, by Theorem 6.4.6, $F(x) - G(x)$ is constant, i.e.

$$\overline{\int_a^x} f\,dt - \underline{\int_a^x} f\,dt \text{ is constant.}$$

But

$$\overline{\int_a^a} f\,dt - \underline{\int_a^a} f\,dt = 0-0 = 0.$$

Therefore

$$\overline{\int_a^x} f\,dt - \underline{\int_a^x} f\,dt = 0$$

and, in particular,

$$\overline{\int_a^b} f\,dt = \underline{\int_a^b} f\,dt.$$

Therefore f is integrable in $[a, b]$. □

The theorem which we have just proved shows that continuity is a criterion for integrability. It also suggests the well-known connection between integration and differentiation for continuous functions. This connection is known as the fundamental theorem of integral calculus, and one of its consequences is a standard method of evaluating integrals.

THEOREM 8.4.4 (the fundamental theorem of integral calculus). *If f is continuous in* $[a, b]$ *and if F is defined, for all x in* $[a, b]$ *by*

$$F(x) = \int_a^x f\,dt$$

then $\qquad F'(x) = f(x)$ *everywhere in* (a, b).

Proof. Since f is continuous it is integrable by Theorem 8.4.3. Therefore

$$\underline{\int_a^b} f\,dt = \overline{\int_a^b} f\,dt.$$

The result now follows immediately from Theorem 8.4.3(i). □

COROLLARY 1. *Every continuous function has an anti-derivative.*

The fundamental theorem gives a method of constructing such a function.

COROLLARY 2. *If G is an anti-derivative of a continuous function f, then*

$$\int_a^b f\,dt = G(b) - G(a).$$

Proof. Let $\qquad F(x) = \int_a^x f\,dt.$

Then F is an anti-derivative of f. Thus, by the corollary to Theorem 6.4.6, F and G differ by a constant, say

$$G(x) = F(x) + c.$$

Thus

$$\begin{aligned} G(b) - G(a) &= F(b) + c - F(a) - c \\ &= \int_a^b f\,dt - \int_a^a f\,dt \\ &= \int_a^b f\,dt. \ \square \end{aligned}$$

Corollary 2 enables us to evaluate many definite integrals as in the following example.

Since $\sin x$ is an anti-derivative of $\cos x$,

$$\int_{\pi/6}^{\pi/2} \cos x\,dx = \sin\frac{\pi}{2} - \sin\frac{\pi}{6} = \frac{1}{2}.$$

Inverse trigonometric functions. The main purpose of inverse trigonometric functions is for the evaluation of integrals. The reader will be able to verify that $\sin^{-1} x/c$ is an anti-derivative of $1/[\sqrt{(c^2-x^2)}]$ and that $(1/c)\tan^{-1}(x/c)\}$ is an anti-derivative of $1/(c^2+x^2)$. Thus

$$\int_a^b \frac{dx}{c^2+x^2} = \frac{1}{c}\left(\tan^{-1}\frac{b}{c}-\tan^{-1}\frac{a}{c}\right)$$

and, provided that $|c| > |a|$ and $|c| > |b|$,

$$\int_a^b \frac{dx}{\sqrt{(c^2-x^2)}} = \sin^{-1}\frac{b}{c}-\sin^{-1}\frac{a}{c}.$$

Area of a circle. In section 7.3, π was defined as twice the lowest positive zero of $\cos x$. In section 8.1 the area of a circle of radius r was shown to be πr^2, where π is the ratio of the lengths of the circumference and the diameter of a circle. We now show that the two definitions of π both provide the same number.

THEOREM 8.4.5. *The area of a circle of radius r is πr^2, where $\pi/2$ is the least positive zero of* $\cos x$, *i.e. the value of* $\cos^{-1}0$ *and* $\sin^{-1} 1$.

Proof. The reader will be able to verify that if f is defined by

$$f(x) = \frac{1}{2}\left(r^2\sin^{-1}\frac{x}{r}+x\sqrt{(r^2-x^2)}\right),$$

then f is an anti-derivative of $\sqrt{(r^2-x^2)}$.

Thus the area of a circle of radius r is

$$4\int_0^r \sqrt{(r^2-x^2)}\,dx$$

$$= 2r^2\sin^{-1}1 = \pi r^2. \quad \square$$

A detailed discussion of the use of inverse trigonometric functions and substitutions in the integration of various function is not immediately relevant to the main theme of this book and is therefore omitted.

Non-continuous integrable functions. It has been shown that, if a function is either continuous or monotonic, then it is integrable. If a function is neither continuous nor monotonic, then it may or may not be integrable.

Consider first the function f defined by

$$\begin{cases} f(x) = 1 & \text{if } 0 \leqslant x < 1, \\ f(x) = 2 & \text{if } 1 \leqslant x < 2, \\ f(x) = \frac{1}{2} & \text{if } 2 \leqslant x \leqslant 3. \end{cases}$$

f has two points of discontinuity in [0, 3] and is not monotonic. However, it is not difficult to show that f is integrable in [0, 3] with

$$\int_0^3 f\,dx = 3\tfrac{1}{2}.$$

Now consider the function f defined by

$$f(x) = 1 \quad \text{if } x \text{ is a rational.}$$
$$f(x) = 0 \quad \text{if } x \text{ is an irrational.}$$

All the intervals of any dissection $\mathcal{D}$ of [0, 1] contain both rational and irrational points. Thus if $S_{\mathcal{D}}$ and $s_{\mathcal{D}}$ are the upper and lower sums of f, corresponding to $\mathcal{D}$, over the interval [0, 1], then $S_{\mathcal{D}} = 1$ and $s_{\mathcal{D}} = 0$.

Therefore $\overline{\int_0^1} f\,dx = 1$ and $\underline{\int_0^1} f\,dx = 0$, and f is not integrable in [0, 1].

Of the two discontinuous functions above, the first is integrable and the second is not. Nevertheless, some kind of pattern is discernible in that the second function is, in an obvious sense, much more discontinuous than the first. To be exact, the first function has only two points of discontinuity, whereas the second is discontinuous everywhere. It can, in fact, be proved that a function with only a finite number of discontinuities is always integrable, while a function discontinuous everywhere is never integrable. The exact criterion in this direction is difficult both to state and to prove, and is beyond the scope of this book.

Approximate methods. The anti-derivative method of evaluating the integral of a continuous function f depends upon the existence of a standard function whose derivative is equal to f. Thus, since the anti-derivative of $\cos x$ is $\sin x$ and since $\sin x$ is a standard function whose value, for different values of x, is obtainable by using published tables, an integral such as

$$\int_{\pi/5}^{4\pi/9} \cos x \, dx$$

may readily be evaluated, thus

$$\int_{\pi/5}^{4\pi/9} \cos x \, dx = \sin \frac{4\pi}{9} - \sin \frac{\pi}{5}$$
$$= 0.9848 - 0.5878 = 0.3970$$

On the other hand, it sometimes happens that, even though a function f is continuous, this method of evaluating the integral of f fails because there is no standard anti-derivative. For example, suppose that we wish to evaluate

$$\int_a^b \log \sin x \, dx.$$

We know, by Corollary 1 to Theorem 8.4.4, that $\log \sin x$ has an anti-derivative. However, this is of little help since the anti-derivative of $\log \sin x$ is not a standard function and there is therefore no set of tables which will give its relevant values.

Nevertheless, if we require a numerical answer, then there are various methods of obtaining this. The simplest is to choose a dissection $\mathcal{D}$ and evaluate $S_{\mathcal{D}}$ and $s_{\mathcal{D}}$. If this is done for the integral above, then we know that

$$s_{\mathcal{D}} \leqslant \int_a^b \log \sin x \, dx \leqslant S_{\mathcal{D}}.$$

If our answer is not sufficiently accurate, then the number of intervals of the dissection may be increased.

There are well-known modifications of this method of approximation, such as Simpson's rule and the trapezoidal rule. The details of the application of these methods are not of interest here.

Exercises 8.4

1. If the function f is differentiable in (a, b), prove that f also has an anti-derivative in (a, b).

2. If the function f is bounded and has only one point of discontinuity in $[a, b]$, prove that f is Riemann integrable in $[a, b]$. Deduce that, if f is bounded and has only a finite number of discontinuities in $[a, b]$, then f is integrable.

3. Give an example of a function which has only one point of discontinuity in $[a, b]$ but which is not integrable in $[a, b]$.

4. In each of the following cases the function f is continuous and therefore has an anti-derivative for all values of x. Find an anti-derivative in each case.

(i) $f(x) = |x|$.

(ii) $\begin{cases} f(x) = x & \text{if } x < 1, \\ f(x) = x^2 & \text{if } x \geqslant 1. \end{cases}$

(iii) $f(x) = |x-2|$.

(iv) $\begin{cases} f(x) = x \log x^2 & \text{if } x \neq 0, \\ f(0) = 0. \end{cases}$

5. Determine whether the non-continuous function f is Riemann integrable in $-1, 1]$ in each of the following cases. Where it is integrable state the value of the ntegral.

(i) $\begin{cases} f(x) = |x|/x & \text{if } x = 0, \\ f(0) = 0. \end{cases}$

(ii) $\begin{cases} f(x) = x & \text{if } x \text{ is rational}, \\ f(x) = 1 & \text{if } x \text{ is irrational}. \end{cases}$

(iii) $\begin{cases} f(x) = x^2 & \text{if } |x| > \frac{1}{2}, \\ f(x) = x & \text{if } |x| \leq \frac{1}{2}. \end{cases}$

6. Show by an example that if the functions f and g are bounded and if the function h is defined by

$$h(x) = f(x) + g(x)$$

for all values of x, then the equation

$$\overline{\int_a^b} h\,dx = \overline{\int_a^b} f\,dx + \overline{\int_a^b} g\,dx$$

does not necessarily hold. (f and g may be chosen to be functions which are discontinuous everywhere). Compare this result with Theorem 8.2.7.

7. If $\mathcal{D}$ is the dissection

$$\left\{\frac{\pi}{6}, \frac{2\pi}{9}, \frac{5\pi}{18}, \frac{\pi}{3}, \frac{7\pi}{18}, \frac{4\pi}{9}, \frac{\pi}{2}\right\}$$

evaluate $S_{\mathcal{D}}$ and $s_{\mathcal{D}}$ for the function $f(x) = \log \sin x$.

Hence find an approximate value for

$$\int_{\pi/6}^{\pi/2} \log \sin x \, dx.$$

8.5. Further applications of the fundamental theorem

Integration of power series. In section 7.1 it was shown that a power series could be differentiated term by term within its circle of convergence. There is a similar result for integration.

THEOREM 8.5.1. *If the function f is defined by*

$$f(x) = \sum_{n=0}^{\infty} a_n x^n,$$

the radius of convergence of the power series being R, and if

$$|a| < R, \quad |b| < R,$$

then

$$\int_a^b f\,dx = \sum_{n=0}^{\infty} a_n \frac{(b^{n+1} - a^{n+1})}{n+1}.$$

Proof. Put $K = \max(|a|, |b|)$. Then $K < R$. Therefore

$$\sum_{n=0}^{\infty} a_n K^n$$

is absolutely convergent. Thus, given $\varepsilon > 0$, we can find a natural number M such that, for all $N \geqslant M$,

$$\sum_{n=N+1}^{\infty} |a_n| K^n < \frac{\varepsilon}{b-a}.$$

Then, if x belongs to $[a, b]$,

$$\begin{aligned}\left|f(x)-\sum_{n=0}^{N} a_n x^n\right| &= \left|\sum_{n=N+1}^{\infty} a_n x^n\right| \\ &\leqslant \sum_{n=N+1}^{\infty} |a_n x^n| \\ &\leqslant \sum_{n=N+1}^{\infty} |a_n| K^n \\ &< \frac{\varepsilon}{b-a}.\end{aligned}$$

Therefore, using Theorem 8.2.9 and its corollary,

$$\begin{aligned}\left|\int_a^b \left(f-\sum_{n=0}^{N} a_n x^n\right) dx\right| &\leqslant \int_a^b \left|f-\sum_{n=0}^{N} a_n x^n\right| dx \\ &\leqslant \int_a^b \frac{\varepsilon}{b-a}\, dx \\ &= \varepsilon.\end{aligned}$$

Thus
$$\left|\int_a^b f\, dx - \int_a^b \sum_{n=0}^{N} a_n x^n\, dx\right| \leqslant \varepsilon$$

or
$$\left|\int_a^b f\, dx - \sum_{n=0}^{N} a_n \frac{b^{n+1}-a^{n+1}}{n+1}\right| \leqslant \varepsilon.$$

Hence
$$\sum_{n=0}^{\infty} a_n \frac{b^{n+1}-a^{n+1}}{n+1} = \int_a^b f\, dx. \quad \square$$

Theorem 8.5.1 may be used to give an alternative and somewhat easier proof that a power series may be differentiated term by term within its circle of convergence (Theorem 7.1.4). This new proof, however, requires that Theorem 7.1.5 be proved independently of Theorem 7.1.4; this is left for the reader.

Theorem 7.1.4 *New proof. Put*

$$g(x) = \sum_{n=0}^{\infty} n a_n x^{n-1}.$$

Let $0 \leqslant X < R$. By Theorem 7.1.5 g is continuous and therefore integrable inside its circle of convergence. Define G by

$$G(X) = \int_0^X g\,dx.$$

By the fundamental theorem, $G'(x) = g(x)$. By Theorem 8.5.1 g is integrable term by term so that

$$\int_0^X g\,dx = \sum_{n=0}^{\infty} a_n X^n = f(X).$$

Thus $G(x) = f(x)$ for all x satisfying $0 \leqslant x < R$. Therefore $f'(x) = g(x)$ as required. □

The proof when $-R < X < 0$ requires a modified form of the fundamental theorem, but is left to the reader (see question 7 of exercises 8.5).

Evaluation of π. Theorem 8.5.1 may also be used to find the Maclaurin series of various functions, inside their circle of convergence, as the following example shows.

If $0 \leqslant x < 1$,

$$1 - x^2 + x^4 - x^6 + x^8 - \ldots$$

is a geometric progression with common ratio $-x^2$. Therefore its sum to infinity is $1/(1+x^2)$. If $0 \leqslant X < 1$, X lies within the circle of convergence of the progression, and Theorem 8.5.1 gives

$$\int_0^X \frac{dx}{1+x^2} = \int_0^X (1 - x^2 + x^4 - x^6 + \ldots)\,dx$$

$$= X - \frac{X^3}{3} + \frac{X^5}{5} - \frac{X^7}{7} + \ldots,$$

that is, $$\tan^{-1} Y = X - \frac{X^3}{3} + \frac{X^5}{5} - \frac{X^7}{7} + \ldots.$$

When $X = 1$ this result may no longer be obtained from Theorem 8.5.1 since 1 does not lie within the circle of convergence of the series. Nevertheless, the reult remains true and gives us a well-known series for π, called Euler's series:

$$\frac{\pi}{4} = 1-\frac{1}{3}+\frac{1}{5}-\frac{1}{7}+\ldots.$$

This is now proved.

$$1-x^2+x^4-x^6+\ldots+(-x^2)^{n-1} = \frac{1-(-x^2)^n}{1+x^2}.$$

Thus $$\left|\frac{1}{1+x^2}-(1-x^2+x^4-\ldots+(-x^2)^{n-1})\right| \leqslant \frac{x^{2n}}{1+x^2} < x^{2n}.$$

Therefore, by Theorem 8.2.9,

$$\left|\int_0^1 \frac{dx}{1+x^2}-(1-x^2+x^4-\ldots+(-x^2)^{n-1})\,dx\right| \leqslant \int_0^1 x^{2n}\,dx = \frac{1}{2n+1}$$

Therefore $$\left|\tan^{-1} 1-\left(1-\frac{1}{3}+\frac{1}{5}-\ldots+\frac{(-1)^{n-1}}{2n-1}\right)\right| \to 0 \text{ as } n \to \infty.$$

Hence $$\frac{\pi}{4} = 1-\frac{1}{3}+\frac{1}{5}-\frac{1}{7}+\ldots.$$

Euler's series provides an infinite series from which π may, in theory, be evaluated. However, the convergence of this series is very slow and, in practice, an inconveniently large number of terms would be required to achieve even modest accuracy. Instead, π may be evaluated using the series for $\tan^{-1} x$, in the manner indicated in question 1 of exercises 8.5.

In the previous chapter, Wallis's infinite product for π was mentioned. This is again too slowly converging to make it of use in evaluating π; nevertheless we now give a derivation of the product.

Put $$I_n = \int_0^{\pi/2} \sin^n x\,dx.$$

Using the method of integration by parts (if the reader is unfamiliar with this, he will find it explained in any textbook of integral calculus), we have

$$I_n = -\cos\frac{\pi}{2}\sin^{n-1}\frac{\pi}{2}+\cos 0\sin^{n-1}0+\int_0^{\pi/2}(n-1)\cos x\sin^{n-2}x$$

$$\times\cos x\,dx = (n-1)(I_{n-2}-I_n)$$

provided that $n \geqslant 2$. Hence, with this proviso,

$$I_n = \frac{n-1}{n}I_{n-2}.$$

But $$I_0 = \frac{\pi}{2},\quad I_1 = 1.$$

Therefore $$I_{2n} = \frac{2n-1}{2n}\cdot\frac{2n-3}{2n-2}\cdots\frac{5}{6}\cdot\frac{3}{4}\cdot\frac{1}{2}\cdot\frac{\pi}{2}$$

and $$I_{2n-1} = \frac{2n-2}{2n-1}\cdot\frac{2n-4}{2n-3}\cdots\frac{6}{7}\cdot\frac{4}{5}\cdot\frac{2}{3}.$$

For all x in $[0, \pi/2]$ $\sin^{n-2}x \geqslant \sin^{n-1}x \geqslant \sin^n x$. Therefore, by the corollary to Theorem 8.2.9,

$$I_{2n-2} \geqslant I_{2n-1} \geqslant I_{2n}.$$

Therefore $$\frac{I_{2n-2}}{I_{2n}} \geqslant \frac{I_{2n-1}}{I_{2n}} \geqslant 1.$$

But $$\frac{I_{2n-2}}{I_{2n}} = \frac{2n}{2n-1}.$$

Thus $$\frac{2n}{2n-1} \geqslant \frac{I_{2n-1}}{I_{2n}} \geqslant 1.$$

Therefore $$\frac{I_{2n-1}}{I_{2n}} \to 1 \text{ as } n \to \infty.$$

Thus $$\lim\left[\frac{2}{3}\cdot\frac{4}{5}\cdot\frac{6}{7}\cdots\frac{2n-2}{2n-1}\cdot\frac{2}{1}\cdot\frac{4}{3}\cdot\frac{6}{5}\cdots\frac{2n}{2n-1}\cdot\frac{2}{\pi}\right] = 1.$$

Hence $\dfrac{\pi}{2} = \lim\left[\dfrac{2}{3}\cdot\dfrac{2}{1}\cdot\dfrac{4}{5}\cdot\dfrac{4}{3}\cdot\dfrac{6}{7}\cdot\dfrac{6}{5}\cdots\dfrac{2n-2}{2n-1}\cdot\dfrac{2n}{2n-1}\right]$

$$= \frac{2}{1}\cdot\frac{2}{3}\cdot\frac{4}{3}\cdot\frac{4}{5}\cdot\frac{6}{5}\cdot\frac{6}{7}\cdot\frac{8}{7}\cdot\frac{8}{9}\cdot\frac{10}{9}\cdot\frac{10}{11}\cdots,$$

which is Wallis's formula.

The evaluation of limits of series. In section 8.2 we saw how the knowledge of sums of certain series enabled important definite integrals to be evaluated.

Now that the fundamental theorem has been proved, the evaluation of a much wider range of integrals is quickly possible and we may reverse the process and deduce the values of certain limits connected with series.

As an example it will be shown that

$$\lim_{n\to\infty}\left[\frac{1}{n^{p+1}}\sum_{r=1}^{n} r^p\right] = \frac{1}{p+1}.$$

Let $\mathcal{D}$ be the dissection

$$\left\{0, \frac{1}{n}, \frac{2}{n}, \ldots, \frac{n}{n}\right\}$$

of $[0, 1]$. If $f(x) = x^p$ and if $S_{\mathcal{D}}$ and $s_{\mathcal{D}}$ are the upper and lower sums corresponding to the function f, then

$$S_{\mathcal{D}} = \sum_{r=1}^{n}\left(\frac{r}{n}\right)^p\frac{1}{n}, \quad s_{\mathcal{D}} = \sum_{r=0}^{n-1}\left(\frac{r}{n}\right)^p\frac{1}{n}.$$

Also
$$\int_0^1 f\,dx = \frac{1}{p+1}.$$

Thus
$$\sum_{r=0}^{n-1}\left(\frac{r}{n}\right)^p\frac{1}{n} \leqslant \int_0^1 f\,dx \leqslant \sum_{r=1}^{n}\left(\frac{r}{n}\right)^p\frac{1}{n}.$$

Rearranging, we have

$$\frac{1}{p+1} \leqslant \sum_{r=1}^{n}\frac{r^p}{n^{p+1}} \leqslant \frac{1}{p+1}+\frac{1}{n}.$$

Hence
$$\sum_{r=1}^{n}\frac{r^p}{n^{p+1}} \to \frac{1}{p+1} \quad \text{as } n\to\infty.$$

Exercises 8.5

1. Use Theorems 7.3.3 and 7.3.4 to prove that

(i) $$\tan(a-b) = \frac{\tan a - \tan b}{1+\tan a \cdot \tan b},$$

(ii) $$\tan(2a) = \frac{2\tan a}{1-\tan^2 a}.$$

Deduce that

$$4\tan^{-1}\frac{1}{5} - \tan^{-1}\frac{1}{239} = \frac{\pi}{4}.$$

Hence evaluate π to three decimal places.

2. Use Theorem 8.5.1 to obtain the log series

$$\log(1+x) = x - \frac{x^2}{2} + \frac{x^3}{3} - \frac{x^4}{4} + \ldots$$

when $|x| < 1$.

3. Use Theorems 7.6.2 and 8.5.1 to obtain the Maclaurin series for $\sin^{-1} x$ up to the term in x^5.

4. Use the integral

$$\int_0^{\pi/2} \sin x \, dx$$

to deduce the value of

$$\lim_{n\to\infty} \left[\frac{1}{n}\sum_{r=1}^{n} \sin\left(\frac{\pi r}{2n}\right)\right].$$

5. Show that

$$\lim_{n\to\infty} \left[\sum_{r=1}^{n} \frac{4n}{(2n-r)^2}\right] = 2.$$

6. Prove Theorem 7.1.5 (without assuming Theorem 7.1.4).

7. If g is continuous in $[a, b]$ and if G is defined by

$$G(x) = \int_x^b g \, dx,$$

show that $G'(x) = -g(x)$.

Hence complete the new proof of Theorem 7.1.4.

8. Prove carefully that the following inequalities hold for all n:

(i) $$1 - \frac{1}{3} + \frac{1}{5} - \frac{1}{7} + \ldots + \frac{1}{4n+1} \geqslant \frac{\pi}{4},$$

(ii) $$1 - \frac{1}{3} + \frac{1}{5} - \frac{1}{7} + \ldots - \frac{1}{4n+3} \leqslant \frac{\pi}{4},$$

8.6. Alternative approach to the logarithmic function

In section 7.4 the logarithmic function was defined as the inverse of the exponential function. In this section an alternative approach is adopted, using the Riemann integral of $1/x$.

THEOREM 8.6.1. *With the logarithmic function as defined in section 7.4,*

$$\int_1^x \frac{dt}{t} = \log x,$$

where $x > 0$.

Proof. By Theorem 7.4.1,

$$\int_1^x \frac{dt}{t} = \log x - \log 1 = \log x. \quad \square$$

In view of Theorem 8.6.1, we may, if we wish, choose to develop the properties of the logarithmic function from the following definition of $\log x$ instead of that given in section 7.4.

DEFINITION. *When* $x > 0$, $\log x = \int_1^x \frac{dt}{t}$.

Most of the properties of the logarithmic function which were proved in section 7.4 will now be proved again from this new standpoint.

THEOREM 8.6.2. *If* $f(x) = \log x$,

(i) *f is a strictly increasing function,*

(ii) $f'(x) = 1/x$.

Proof. (i) If $b > a$,

$$f(b) - f(a) = \int_1^b \frac{dt}{t} - \int_1^a \frac{dt}{t}$$

$$= \int_a^b \frac{dt}{t} \geqslant \frac{b-a}{b} > 0.$$

(ii) Since $1/x$ is continuous, the result is the immediate consequence of the definition of $\log x$ and the fundamental theorem of calculus.

THEOREM 8.6.3. *If a and b are positive,*

$$\log ab = \log a + \log b.$$

Proof. Let $F(x) = \log ax$. Then

$$F'(x) = \frac{a}{ax} = \frac{1}{x},$$

i.e. F is an anti-derivative of $1/x$. Thus

$$\log b = \int_1^b \frac{dt}{t} = F(b) - F(1)$$

$$= \log ab - \log a.$$

THEOREM 8.6.4. (i) $\log 1/x = -\log x$.

(ii) $\log x \to \infty$ as $x \to \infty$.

(iii) $\log x \to \infty$ as $x \to 0+0$.

(iv) $$\lim_{x \to 0} \left[\frac{\log (1+x)}{x} \right] = 1.$$

Proof. (i) $\log 1/x + \log x = \log \{(1/x) \cdot x\} = \log 1 = 0.$

(ii) $\log 2 > 0. \qquad \log 2^n = n \cdot \log 2.$

Given any real number K, we can find a natural number N with $N \cdot \log 2 > K$.

If $x \geqslant N$, $\qquad \log x \geqslant \log 2^N = N \cdot \log 2 > K.$

Thus $$\log x \to \infty \quad \text{as} \quad x \to \infty.$$

(iii) $$\log x \to \infty \quad \text{as} \quad x \to \infty.$$

Therefore $$-\log x \to -\infty \quad \text{as} \quad x \to \infty.$$

Thus $$\log 1/x \to -\infty \quad \text{as} \quad x \to \infty.$$

Hence $$\log x \to -\infty \quad \text{as} \quad x \to 0+0.$$

(iv) $$\log (1+x) = \int_1^{1+x} \frac{dt}{t}.$$

Therefore, if $x > 0$, Theorem 8.4.1 gives

$$\frac{x}{1+x} \leqslant \log (1+x) \leqslant x.$$

Thus $$\frac{1}{1+x} \leqslant \frac{\log (1+x)}{x} \leqslant 1.$$

Therefore $$\frac{\log (1+x)}{x} \to 1 \quad \text{as} \quad x \to 0+0.$$

If $x < 0$, Theorem 8.4.1 gives

$$-x \leqslant -\log (1+x) \leqslant \frac{-x}{1+x}.$$

Dividing by $-x$, which is positive, this gives

$$1 \leqslant \frac{\log (1+x)}{x} \leqslant \frac{1}{1+x}.$$

Therefore $$\frac{\log (1+x)}{x} \to 1 \quad \text{as} \quad x \to 0-0.$$

Thus $$\lim \frac{\log (1+x)}{x} = 1 \,\square.$$

THEOREM 8.6.5. *If a is a positive real number, then*

$$\frac{\log x}{x^a} \to 0 \quad as \quad x \to \infty .$$

Proof. This theorem is proved by showing that, if $2b = a$, then

$$\frac{\log x}{x^a} = \frac{\log x}{x^b} \cdot \frac{1}{x^b} < \frac{1}{bx^b} .$$

The details are left to the reader (see question 5 of exercises 8.6).

The logarithmic function and the harmonic series. There is a remarkable connection between the logarithmic function and the harmonic series, a connection which enables us to estimate the size of the first n terms of the harmonic series.

THEOREM 8.6.6. *For all natural numbers n,*

$$0 \leqslant \sum_{r=0}^{n} \frac{1}{r} - \log n \leqslant 1.$$

Proof.
$$\log n = \int_1^n \frac{dx}{x} .$$

Let $\mathscr{D}$-be the dissection

$$\{1, 2, 3, 4, \ldots, n\}$$

of the interval $[1, n]$. If $S_{\mathscr{D}}$ and $s_{\mathscr{D}}$ are the upper and lower sums corresponding to the function $1/x$, then

$$S_{\mathscr{D}} = 1 + \frac{1}{2} + \frac{1}{3} + \frac{1}{4} + \ldots + \frac{1}{n-1} ,$$

$$s_{\mathscr{D}} = \frac{1}{2} + \frac{1}{3} + \frac{1}{4} + \frac{1}{5} + \ldots + \frac{1}{n} .$$

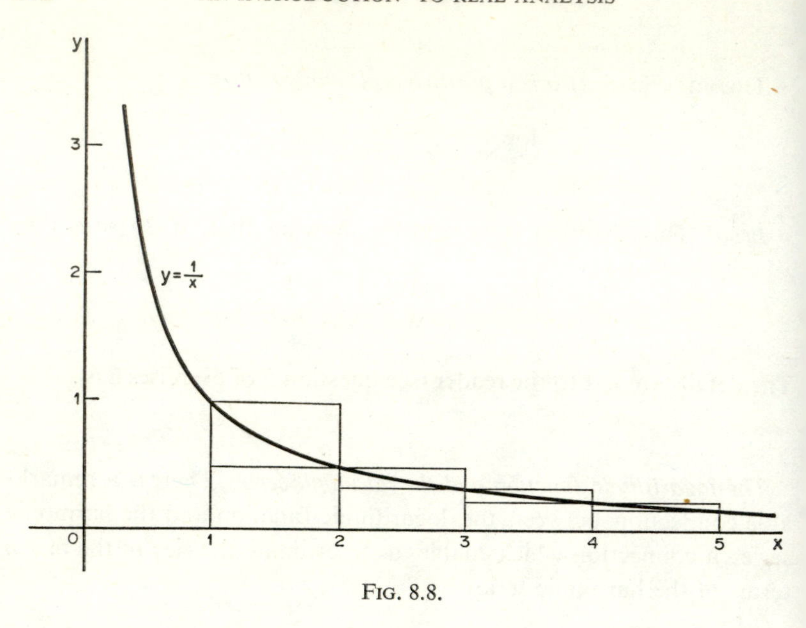

FIG. 8.8.

Figure 8.8 indicates the situation when $n = 5$.

$$s_{\mathcal{D}} \leqslant \log n \leqslant S_{\mathcal{D}}.$$

Therefore

$$\sum_{r=1}^{n} \frac{1}{r} - 1 \leqslant \log n \leqslant \sum_{r=1}^{n} \frac{1}{r}.$$

Hence

$$0 \leqslant \sum_{r=1}^{n} \frac{1}{r} - \log n \leqslant 1. \square$$

In fact we may take the matter a stage further.

THEOREM 8.6.7. $\sum_{r=1}^{n} 1/r - \log n \to \gamma$ *as* $n \to \infty$, *where* γ *is a constant, known as Euler's constant, and satisfies*

$$0 \leqslant \gamma \leqslant 1.$$

Proof.

$$\left(\sum_{r+1}^{n+1}\frac{1}{r}-\log(n+1)\right)-\left(\sum_{r=1}^{n}\frac{1}{r}-\log n\right)$$

$$= \frac{1}{n+1}-(\log(n+1)-\log n)$$

$$= \frac{1}{n+1}-\int_{n}^{n+1}\frac{dx}{x}$$

$$\leqslant \frac{1}{n+1}-\frac{1}{n+1} \quad \text{by Theorem 8.4.1}$$

$$= 0.$$

Thus $1/r-\log n$ is a decreasing function of n. It is bounded below by zero, by Theorem 8.6.6 Therefore it tends to a limit γ, which satisfies $0\leqslant\gamma\leqslant 1$ in view of Theorem 8.6.6.

The anti-derivative of $1/x$. If $x>0$, one anti-derivative of $1/x$ is $\log x$. (As we have seen, $\log ax$ is also an anti-derivative for any positive constant a.)

If $x<0$, $\log x$ is not defined in real analysis. However, if $f(x) = \log(-x)$, then $f'(x) = -1/-x = 1/x$. Thus $\log|x|$ is an anti-derivative of $1/x$ for all non-zero x.

We may use this result to evaluate definite integrals. For example,

$$\int_{-4}^{-2}\frac{dx}{x} = \log|-2|-\log|-4|$$

$$= \log 2-\log 4$$

$$= -\log 2.$$

The reader may note that

$$\int_{-4}^{-2}\frac{dx}{x} = \log(-2)-\log(-4)$$

$$= \log\frac{-2}{-4}$$

$$= \log\tfrac{1}{2} = -\log 2$$

is a nonsense, since $\log -2$ and $\log -4$ do not exist in real analysis.

Exercises 8.6

1. Use tables to evaluate

$$\text{(i)} \int_{2}^{7} \frac{3}{x}\,dx, \quad \text{(ii)} \int_{-5}^{-3} \frac{2}{x}\,dx$$

correct to two decimal places.

2. Comment upon the following:

$$\int_{-2}^{3} \frac{dx}{x} = \log 3 - \log|-2| = \log 3 - \log 2 = \log 1\tfrac{1}{2}.$$

3. Modify Theorem 8.6.6 to prove that

$$0 < \gamma < 1.$$

4. Use a series for evaluating logarithms to show that

$$\log 3 > 1.095.$$

Deduce an inequality for log 9 and hence show that

$$\gamma < 0.64.$$

5. If $b > 0$, show that

$$x^b = b\int_{1}^{x} \frac{dt}{t^{1-b}} + 1.$$

Hence, using the corollary to Theorem 8.2.9, show that when $x > 1$,

$$f(x) = \frac{\log x}{x^a} < \frac{1}{bx^b},$$

where $a = 2b$. Deduce that $f(x) \to 0$ as $x \to \infty$.

6. If $f(x) = x \cdot \log x - x$, show that $f'(x) = \log x$.

Evaluate

$$\int_{1}^{n} \log x\,dx.$$

Hence show that

$$\sum_{r=1}^{n-1} \log r \leqslant n \cdot \log n - n + 1 \leqslant \sum_{r=2}^{n} \log r.$$

Deduce the limit, as n tends to infinity, of

$$\frac{1}{(n+1)!}\left(\frac{n}{e}\right)^n.$$

7. Theorem 8.6.7 may be expressed as follows:

$$\sum_{r=1}^{n} \frac{1}{r} = \log n + R_n,$$

where $R_n \to \gamma$ as $n \to \infty$. Use this result to find an expression for the sum of the first $2n$ terms of the series

$$1-\tfrac{1}{2}+\tfrac{1}{3}-\tfrac{1}{4}+\tfrac{1}{5}-\tfrac{1}{6}+\tfrac{1}{7}-\ldots$$

and deduce the sum to infinity of this series.

8. If the series of question 7 is rearranged so that two positive terms are taken for every negative term, thus,

$$1+\tfrac{1}{3}-\tfrac{1}{2}+\tfrac{1}{5}+\tfrac{1}{7}-\tfrac{1}{4}+\tfrac{1}{9}+\tfrac{1}{11}-\tfrac{1}{6}+\ldots,$$

find an expression (as in question 7) for the sum of the first $3n$ terms of this series. Deduce that the series is convergent and that its sum to infinity is $\frac{1}{2}\cdot\log 8$.

9. If the series of question 7 is rearranged so that we take one positive term, then one negative, then three positive, then one negative, then five positive, then one negative, then seven positive and so on, thus,

$$1-\tfrac{1}{2}+\tfrac{1}{3}+\tfrac{1}{5}+\tfrac{1}{7}-\tfrac{1}{4}+\tfrac{1}{9}+\tfrac{1}{11}+\tfrac{1}{13}+\tfrac{1}{15}+\tfrac{1}{17}-\tfrac{1}{6}+\ldots,$$

find an expression (as in question 7) for the sum of the first n^2+n terms of this series. Deduce that this series is divergent.

10. How many terms of the harmonic series must be taken in order to ensure that their sum exceeds 30?

8.7. Infinite and improper integrals

The Riemann integral is defined over a closed interval $[a, b]$ for a function f which is bounded in $[a, b]$. It is sometimes possible to waive one (or both) of these restrictions and to give meaning to the integral of a function over a semi-infinite interval of the form $[a, \infty)$ or $(-\infty, a]$, or the infinite interval $(-\infty, \infty)$, or to give meaning to the integral of a function which is not bounded in $[a, b]$.

DEFINITION. *The expression*

$$\int_a^\infty f\,dx$$

is called an infinite integral. If

$$\lim_{X\to\infty}\left[\int_a^X f\,dx\right]$$

exists and is finite, then the infinite integral is said to be convergent and its value is the value of the limit.

DEFINITION. *The expression*

$$\int_{-\infty}^{a} f\,dx$$

is called an infinite integral. If

$$\lim_{X\to-\infty}\left[\int_X^a f\,dx\right]$$

exists and is finite, then the infinite integral is said to be convergent and its value is the value of the limit.

DEFINITION. *The infinite integral*

$$\int_{-\infty}^{\infty} f\,dx$$

is said to be convergent if both of the integrals

$$\int_0^\infty f\,dx \quad \textit{and} \quad \int_{-\infty}^{0} f\,dx$$

are convergent and its value is then the sum of the values of the other two integrals.

We illustrate these definitions by means of the following examples:

(1) $$\int_1^X \frac{dx}{x^2} = 1-\frac{1}{X}.$$

$$\lim_{X\to\infty}\left[1-\frac{1}{X}\right] = 1.$$

Therefore $\int_1^\infty \frac{dx}{x^2}$ is convergent and $\int_1^\infty \frac{dx}{x^2} = 1.$

(2)
$$\int_1^X \frac{dx}{x} = \log X.$$

$$\log X \to \infty \quad \text{as} \quad X \to \infty.$$

Therefore $\int_1^\infty \frac{dx}{x}$ is divergent.

(3)
$$\int_X^0 \exp x\, dx = 1 - \exp X.$$

$$\lim_{X \to -\infty} [1 - \exp X] = 1.$$

Therefore $\int_{-\infty}^0 \exp x\, dx$ is convergent and $\int_{-\infty}^0 \exp x\, dx = 1.$

$$\int_0^X \exp x\, dx = \exp X - 1.$$

$$\exp X - 1 \to \infty \quad \text{as} \quad X \to \infty.$$

Thus neither

$$\int_0^\infty \exp x\, dx \quad \text{nor} \quad \int_{-\infty}^\infty \exp x\, dx$$

are convergent.

DEFINITION. *Suppose that the function f is unbounded in the interval $[a, c]$ but that, given any δ satisfying*

$$0 < \delta < c - a,$$

it is bounded in the interval $[a, c - \delta]$.

Then the expression

$$\int_a^c f\, dx$$

is called an improper integral. If

$$\lim_{\delta \to 0} \left[\int_a^{c-\delta} f\, dx \right]$$

exists and is finite, then the improper integral is said to be convergent and its value is the value of the limit.

DEFINITION. *Suppose that the function f is unbounded in the interval $[c, b]$, but that given any δ satisfying*

$$0 < \delta < b-c,$$

it is bounded in the interval $[c+\delta, b]$.

Then the expression

$$\int_c^b f\, dx$$

is called an improper integral. If

$$\lim_{\delta \to 0} \left[\int_{c+\delta}^{b} f\, dx \right]$$

exists and is finite, then the improper integral is said to be convergent and its value is the value of the limit.

If the improper integral

$$\int_a^c f\, dx$$

and the improper integral

$$\int_c^b f\, dx$$

are both convergent then the improper integral

$$\int_a^b f\, dx$$

is said to be convergent and its value is the sum of the values of the other two integrals. In a similar manner we can deal with the improper integral

$$\int_a^b f\,dx$$

when there are a finite number of points in the interval $[a, b]$ *at which* f *is unbounded.*

These definitions are illustrated by means of the following examples:

(1)
$$\int_\delta^4 \frac{dx}{\sqrt{x}} = 2\sqrt{4}-2\sqrt{\delta},$$
$$\lim_{\delta\to 0}[2\sqrt{4}-2\sqrt{\delta}] = 4.$$

Therefore $\int_0^4 \frac{dx}{\sqrt{x}}$ is convergent and $\int_0^4 \frac{dx}{\sqrt{x}} = 4.$

(2)
$$\int_{1+\delta}^2 \frac{dx}{\sqrt[3]{(x-1)}} = \frac{3}{2}(1-\delta^{2/3}),$$
$$\lim_{\delta\to 0}\left[\frac{3}{2}(1-\delta^{2/3})\right] = \frac{3}{2}.$$
$$\int_0^{1-\delta} \frac{dx}{\sqrt[3]{(x-1)}} = \frac{3}{2}(\delta^{2/3}-1),$$
$$\lim_{\delta\to 0}\left[\frac{3}{2}(\delta^{2/3}-1)\right] = -\frac{3}{2}.$$

Therefore $\int_0^2 \frac{dx}{\sqrt[3]{(x-1)}}$ is convergent and $\int_0^2 \frac{dx}{\sqrt[3]{(x-1)}} = 0.$

The following theorem makes it possible to deduce the convergence of a series from the convergence of an infinite integral and vice versa.

THEOREM 8.7.1 (the comparison test for series and integrals). *If* f *is a decreasing function (whose domain includes the interval* $[1, \infty)$*)*

and if $f(x) \geqslant 0$ *for all* x, *then*

$$\int_1^\infty f\,dx$$

is convergent if and only if

$$\sum_{r=1}^{\infty} f(r)$$

is convergent.

Proof. Let $\mathcal{D}$ be the dissection

$$\{1, 2, 3, \ldots, n\}$$

of the interval $[1, n]$. Let $S_{\mathcal{D}}$ and $s_{\mathcal{D}}$ be the upper and lower sums corresponding to the function f. Then

$$S_{\mathcal{D}} = \sum_{r=1}^{n-1} f(r), \qquad s_{\mathcal{D}} = \sum_{r=2}^{n} f(r),$$

since f is a decreasing function. Therefore

$$\sum_{r=2}^{n} f(r) \leqslant \int_1^n f\,dx \leqslant \sum_{r=1}^{n-1} f(r).$$

Rearranging, this gives

$$0 \leqslant f(n) \leqslant \sum_{r=1}^{n} f(r) - \int_1^n f\,dx \leqslant f(1).$$

Also
$$\left(\sum_{r=1}^{n+1} f(r) - \int_1^{n+1} f\,dx\right) - \left(\sum_{r=1}^{n} f(r) - \int_1^n f\,dx\right)$$

$$= f(n+1) - \int_n^{n+1} f\,dx \leqslant 0,$$

since f is a decreasing function. Thus

$$\sum_{r=1}^{n} f(r) - \int_1^n f\,dx$$

is a decreasing function of n, bounded below by zero and hence tends to a limit as n tends to infinity.

If $\int_1^\infty f\,dx$ is convergent, then $\int_1^n f\,dx$ tends to a limit as n tends to infinity, so that $\sum_{r=1}^{n} f(r)$ tends to a limit as n tends to infinity and $\sum_{r=1}^{\infty} f(r)$ is convergent.

Conversely, suppose $\sum_{r=1}^{n} f(r)$ tends to a limit as n tends to infinity. Then so does $\int_1^n f\,dx$. Let this limit be l.

Since $\int_1^n f\,dx$ is an increasing sequence, l is an upper bound. Since $\int_1^X f\,dx$ is an increasing function of X, l is an upper bound of this function. Thus

$$\int_1^X f\,dx$$

is increasing and bounded above. Hence it tends to a limit and

$$\int_1^\infty f\,dx$$

is convergent. □

Exercises 8.7

1. Say which of the following infinite or improper integrals are convergent and, where they are convergent, evaluate them:

(i) $\int_3^\infty \frac{dx}{x}$.

(ii) $\int_0^\infty x \cdot e^{-x}\,dx$.

(iii) $\int_{-2}^2 \frac{dx}{x^3}$.

(iv) $\int_{-2}^2 \frac{dx}{x}$.

(v) $\int_1^\infty \frac{dx}{x^2}$.

(vi) $\int_{-\infty}^\infty x \cdot e^{-x^2}\,dx$.

2. Show that

$$\int_0^2 \frac{dx}{1-x}$$

is not convergent, although

$$\lim_{\delta\to 0}\left[\int_0^{1-\delta}\frac{dx}{1-x}+\int_{1+\delta}^{2}\frac{dx}{1-x}\right] \quad \text{is finite.}$$

3. Prove that

$$\int_1^\infty \frac{\sin x\, dx}{x^2}$$

is convergent.

4. Use Theorem 8.7.1 to show that

(i) $\displaystyle\sum_{n=2}^{\infty}\frac{1}{n\cdot\log n}$ is divergent.

(ii) $\displaystyle\sum_{n=2}^{\infty}\frac{1}{n(\log n)^a}$ is convergent if $a > 1$ and divergent if $a < 1$.

(iii) $\displaystyle\sum_{n=2}^{\infty}\frac{1}{n\cdot\log n\log(\log n)}$ is divergent.

5. Show that the following are convergent:

(i) $\displaystyle\sum_{n=0}^{\infty} e^{-n}$. (ii) $\displaystyle\int_0^\infty \frac{x^9\, dx}{2^x}$.

8.8 Volumes of revolution

This chapter is concluded by a brief discussion of the measurement of volume.

Volume is measured in units consisting of cubes. In the same way that the appropriate area for a rectangle whose sides are the real numbers x and y is seen to be xy, it may be shown that the appropriate volume for a rectangular prism whose dimensions are the real numbers x, y, and z is xyz. After this it is possible to ascribe a meaning to the volume of a parallelepiped, a tetrahedron, and, finally, any polyhedron.

We can now give a definition of the volume of any set A of points in three-dimensional space.

Let S be the set of all polyhedra containing A and let T be the set of all polyhedra contained in A. Let k be the least upper bound of the volumes of members of T and let K be the greatest lower bound of the volumes of members of S. Then $k \leqslant K$. If $k = K$, then we say that the set A is measurable and that the volume of A is k.

Volume of a circular disc. By an argument similar to that of Theorem 8.1.1 it is shown that the volume of a circular disc of radius r and thickness h is $\pi r^2 h$.

Volume of a solid of revolution. The problem of finding the volume of a solid of revolution may be solved by integration, as indicated in the following theorem.

THEOREM 8.8.1. *Suppose that the area bounded by the curve $y = f(x)$ and the lines $x = a$, $x = b$, $y = 0$ is rotated about the x-axis. Then the volume* of the solid formed is

$$\pi \int_a^b g\, dx,$$

where g is the function defined by

$$g(x) = [f(x)]^2.$$

Proof. Let $\mathcal{D}$ be a dissection of $[a, b]$. Suppose that the solid is cut by planes perpendicular to the x-axis at positions corresponding to the points of $\mathcal{D}$. Also suppose that there are discs inscribed in and circumscribing the solid between each pair of planes. Then the solid will contain a solid consisting of discs, whose volume is

$$\pi\, s_{\mathcal{D}}(g)$$

and will be contained in a solid consisting of discs, whose volume is

$$\pi S_{\mathcal{D}}(g).$$

Since this is true of all dissections $\mathcal{D}$, the volume of the solid must be

$$\pi \int_a^b g\,dx$$

provided that this integral exists. □

Exercises for Chapter 8

1. State Rolle's theorem.

By considering the function

$$F(x) = (b-a)\int_a^x f(t)\,dt-(x-a)\int_a^b f(t)\,dt,$$

where f is continuous in the closed interval $[a, b]$, show that there is a point c in the open interval (a, b) such that $(b-a)f(c) = \int_a^b f(x)\,dx$. State clearly any properties of integrals you use.

Find $$\lim_{x\to 0}\frac{1}{x}\int_1^{1+x}\log t\,dt.$$ [T.C.]

2. Define the lower and upper Riemann integrals of a bounded function $f(x)$ over an interval $[a, b]$. State a set of conditions on $f(x)$ to ensure that $\int_a^b f(x)\,dx$ exists.

Show that the function $f(x) = x \sin x$ is integrable over any interval $[a, b]$, stating clearly any properties of functions or integrals which you use.

By evaluating $\int_a^b x \sin x\,dx$ over a suitable interval, show that

$$S_1 = \lim_{n\to\infty}\sum_{r=1}^{n-1}\frac{r}{n^2}\sin\frac{r\pi}{n} = \frac{1}{\pi}.$$

Evaluate $$S_2 = \lim_{n\to\infty}\sum_{r=1}^{n-1}\frac{r}{n^2}\cos\frac{r\pi}{n}.$$ [T.C.]

3. State carefully what you mean by the Riemann integral

$$\int_a^b u(x)\,dx$$

of the function u over the interval $[a, b]$. Explain why neither of the functions f and g below possesses a Riemann integral over the interval [0, 1]. Explain also why h is integrable over [0, 1].

$$f(x) = \begin{cases} x^{-1} & \text{for } x \text{ real and non-zero,} \\ 0 & \text{for } x = 0. \end{cases}$$

$$g(x) = \begin{cases} x & \text{for } x \text{ rational,} \\ 1-x & \text{for } x \text{ irrational.} \end{cases}$$

$$h(x) = \frac{1}{\sqrt{(4-x^2)}} = (|x| < 2).$$

By using the function h, show that

$$\lim_{n\to\infty} \sum_{r=1}^{n} \frac{1}{\sqrt{(4n^2-r^2)}} = \frac{\pi}{6}. \qquad \text{[T.C].}$$

4. Define the upper and lower Riemann integrals of a bounded function over a closed interval. Show that they are equal if the function is continuous over this interval, stating, without proof, any properties of these integrals and continuous functions that you use.

Form the upper sum for the function $1/(1+x^2)$ over the closed interval [0,1] with a division of the interval into n equal subintervals. Hence show that

$$\lim_{n\to\infty} \sum_{r=0}^{n} \frac{n}{n^2+r^2} = \frac{\pi}{4}. \qquad \text{[T.C.]}$$

5. (i) Prove that if, as n tends to infinity, the sequence $\{s_n\}$ has limit s, and the sequence $\{t_n\}$ has limit t, then the sequence $\{s_n t_n\}$ has limit st. State the corresponding result for the quotient sequence $\{s_n/t_n\}$.

(ii) Give an example to show that $\{s_n t_n\}$ may converge, even if $\{s_n\}$ and $\{t_n\}$ both diverge.

(iii) Show, by considering rectangles of unit width under the curve $y = f(x)$, that if $f(x)$ is continuous and monotonic decreasing, and if $f(x) > 0$, then

$$\int_1^n f(x)\,dx > f(2)+f(3)+\ldots+f(n).$$

Denoting the integral by I_n, and the sum $f(1)+f(2)+\ldots+f(n)$ by S_n, show that

$$f(1) > S_n - I_n > 0.$$

By considering $(S_{n+1}-I_{n+1})-(S_n-I_n)$, show that S_n-I_n is monotonic decreasing.
[T.C.]

6. The functions log and exp are defined by

(a) $$\log x = \int_1^x \frac{dt}{t} \ (x > 0),$$

(b) $$\exp u = x \quad \text{if and only if} \quad u = \log x.$$

Using these definitions prove the following results.

(i) $$\log \frac{p}{q} = \log p - \log q \quad (p > 0, \quad q > 0).$$

(ii) $$\frac{\exp a}{\exp b} = \exp(a-b),$$

(iii) $$0 < \log x < \frac{x^m - 1}{m} \quad (m > 0, \quad x > 1).$$

[B.Ed.]

7. (a) Defining the logarithmic function of the real variable x by

$$\log x = \int_1^x \frac{1}{t} \quad dt \quad (x > 0),$$

prove that, if n is a positive integer, then

(i) $$\log x^n = n \log x,$$

(ii) $$0 < n \log x < x^n \quad \text{for} \quad x > 1,$$

(iii) $$\frac{\log x}{x^n} \to 0 \quad \text{as} \quad x \to \infty.$$

(b) By assuming the following inequality, or otherwise, prove that e is not rational:

$$1+\frac{1}{1!}+\frac{1}{2!}+\frac{1}{3!}+\ldots+\frac{1}{n!} < e < 1+\frac{1}{1!}+\frac{1}{2!}+\frac{1}{3!}+\ldots+\frac{1}{n!}+\frac{1}{n!\,n}.$$

[B.Ed.]

8. Show that, for all values of n,

$$1+\frac{1}{1!}+\frac{1}{2!}+\frac{1}{3!}+\ldots+\frac{1}{n!} < e < 1+\frac{1}{1!}+\frac{1}{2!}+\frac{1}{3!}+\ldots+\frac{1}{n!}+\frac{1}{n!\,n}.$$

9. A function $f(x)$ is bounded when $0 \leqslant x \leqslant 1$; define its upper and lower Riemann integrals over [0, 1] and show by an example that they need not be equal.

Suppose that f is monotonic decreasing and that S_n, s_n are its upper and lower sums respectively corresponding to the subdivision 0, $1/n$, $2/n$, ..., r/n..., 1 (n any positive integer). Prove that

$$0 \leqslant S_n - s_n \leqslant \{f(0)-f(1)\}/n.$$

By taking $f(x) = 1/(x+1)$, show that

$$0 < \frac{1}{n}+\frac{1}{n+1}+\frac{1}{n+2}+\ldots+\frac{1}{2n-1}-\log 2 < \frac{1}{2n}.$$

(It may be assumed that the Riemann integral of $F'(x)$ over [0, 1], if it exists, is equal to $F(1)-F(0)$). [B.Ed.]

10. Given that $f(x)$ is continuous in the closed interval $a \leqslant x \leqslant b$, prove that, for some t in the open interval $a < t < b$,

$$\int_a^b f(u)\,du = (b-a)f(t),$$

the integral being defined in the Riemann sense.

(i) If $F(x)$ is defined by

$$F(x) = \int_a^x f(u)\,du$$

prove that

$$\lim_{h \to 0}\left\{\frac{F(x+h)-F(x)}{h}\right\} = f(x) \quad (a \leqslant x+h \leqslant b)$$

by applying the above mean value theorem.

(ii) Define and evaluate the improper integral

$$\int_{-1}^{1} \frac{1}{\sqrt{(1-x^2)}}\,dx.$$ [B.Ed.]

11. Define the upper integral $\overline{\int_a^b} f\,dx$ of a function f which is defined and bounded in $[a, b]$.

Prove that, if $m < f(x) < M$ in $[a, b]$, then

$$(b-a)m < \overline{\int_a^b} f\,dx < (b-a)M.$$

(i) Assuming that

$$\overline{\int_a^c} f\,dx + \overline{\int_c^b} f\,dx = \overline{\int_a^b} f\,dx \quad \text{for} \quad a < c < b,$$

prove that at points x, where f is continuous, $\overline{\int_a^x} f\,dt$ is differentiable as a function of x, and that

$$\frac{d}{dx}\overline{\int_a^x} f\,dt = f(x).$$

(ii) Using (i), or otherwise, obtain the corresponding results for the lower integral $\underline{\int_a^x} f\,dt$, and by considering

$$\frac{d}{dx}\left[\overline{\int_a^x} - \underline{\int_a^x}\right]$$

prove that, if f is continuous in $[a, b]$, then it is integrable there. [B.Ed.]

12. By considering the integral $\int_1^n x^{-1/2}\,dx$, or otherwise, prove that

$$2\sqrt{n}-2+\frac{1}{\sqrt{n}} \leqslant 1+\frac{1}{\sqrt{2}}+\frac{1}{\sqrt{3}}+\ldots+\frac{1}{\sqrt{n}} \leqslant 2\sqrt{n}-1. \qquad \text{[B.Sc.]}$$

13. Define the exponential function e^x and the logarithmic function $\log x$ and show from your definitions that

$$\lim_{x\to 0}\frac{1}{x}\log(1+x) = 1$$

and that

$$\lim_{x\to\infty}\frac{x^k}{e^x} = 0$$

for any positive integer k.

Find

(i) $\lim_{n\to\infty}\left(1+\sin\frac{a}{n}\right)^n$, (ii) $\lim_{y\to 0}(\cos y)^{1/y^2}$. [B.Sc.]

14. A student, asked to define $\log x$, wrote down

$$\log x = \int \frac{1}{x}\,dx.$$

Criticize this answer and give a correct version of the definition.

By examining the graph of the function $1/x$ or otherwise prove that

$$\frac{1}{n+1} < \int_n^{n+1}\frac{1}{x}\,dx < \frac{1}{n}$$

for every positive integer n. By summing these inequalities over a suitable range of values of n, prove that

$$\frac{1}{N} < 1+\frac{1}{2}+\ldots+\frac{1}{N}-\log N < 1$$

for any integer $N \geqslant 2$. [B.Sc.]

15. If the function $f(x)$ is positive and decreases as x increases, show that

$$\int_r^{r+1} f(x)\,dx \leqslant f(r) \leqslant \int_r^{r-1} f(x)\,dx,$$

where r is a positive integer.

If

$$a_n = \frac{n}{n^2+1^2}+\frac{n}{n^2+2^2}+\ldots+\frac{n}{n^2+n^2},$$

where n is a positive integer, show that $a_n \to \pi/4$ as $n \to \infty$. [B.Sc.]

16. If $f(x)$ is a positive monotonic decreasing function of x for $x \geqslant 1$, and if

$$s_n = \sum_{r=1}^{n} f(r), \quad I_n = \int_1^n f(x)\,dx,$$

prove that $s_n - I_n$ tends to a finite limit as $n \to \infty$. Hence prove that the series $\sum_1^\infty f(n)$ and the integral $\int_1^\infty f(x)\,dx$ are either both convergent or both divergent.

Prove that the series

$$\frac{1}{1^a}+\frac{1}{2^b}+\frac{1}{3^a}+\frac{1}{4^b}+\ldots+\frac{1}{(2n-1)^a}+\frac{1}{(2n)^b}+\ldots$$

converges if and only if both a and b are greater than 1. [B.Sc.]

ANSWERS AND HINTS

Exercises for Chapter 1

1. (b) and (d) are not functions

2. (a) $x \to 2x^3-1$, $x \to (2x-1)^3$, $x \to 2(2x-1)-1$.

(b) $x \to \frac{1}{2}(x+1)$, $x \to \sqrt[3]{(\frac{1}{2}x+1)}$.

(c) 1, 0.

3. (b) and (c) are.

Exercises 2.1

1. 3, 4, 4, 6, 7.

2. $3 < 17 < 61 < 145 < 211 < 891 < 6573$.

3. (i) $9-6 = 3$. (ii) $40 \div 8 = 5$. (iii) $81 \div 9 = 9$.

4. Distributive over addition.

5. Yes.

Exercises 2.2

2. A least member.

Exercises 2.3

5. No.

Exercises 2.4

2. -1, $1\frac{7}{10}$, $1\frac{29}{40}$, $-1\frac{7}{8}$ all belong to L; 2, $1\frac{3}{4}$ belong to R.

5. E.g. 2, $2\frac{1}{2}$, π; 1, 0.

6. E.g. $\{a : 0 < a < 1\}$.

7. *Hint:* let $2-a^2 = r$. Then

$$\left(a+\frac{r}{8}\right)^2 = a^2+\frac{ar}{4}+\frac{r^2}{64} < a^2+\frac{r}{2}+\frac{r}{32} < a^2+r = 2.$$

$\sqrt{2}$, $-\sqrt{2}$.

9. Not complete.

10. Not a field.

EXERCISES 2.5

5. E.g. $3+\sqrt{2}$, $3-\sqrt{2}$.

7. *Hint:* consider $r+s-r/\sqrt{2}$.

EXERCISES 2.6

3. If $x = \sqrt{2}$, $y = -\sqrt{2}$, then zero does not belong either to L' or R'.

8. *Hint:* Suppose that $(p/n)^2 < 2$ and that $((p+1)/n)^2 > 2$. Then $p < 2n$. Also $((p+1)/n)^2-(p/n)^2 = (2p+1)/n^2 < 5/n$. We may make this smaller than $2-c$, by taking n sufficiently large.

EXERCISES 3.1

1. (i)n^3. (ii) $1/2^{n-1}$. (iii) $(-1)^{n+1}\,n$. (iv) $2-1/2^{n-1}$. (v) $n/2(1-(-1)^n)$.
(vi) $a_{3n} = 1$, $a_{3n+1} = -1$, $a_{3n+2} = 0$. (vii) $n+(-1)^{n+1}/n$.

2. (i), (ii), (iii) are strictly increasing.
(iv), (vi) are strictly decreasing.
(viii), (ix) are increasing.
(ix), (x) are decreasing.

3. (i) Yes. (ii) No; e.g. consider $a_n = b_n = -n$.

EXERCISES 3.2

1. $2+1/3^{n-1}$, 2. $N = 8$.

2. (i) ∞. (ii) ∞. (iii) 0. (iv) 0.

3. Zero for $-1 < x < 1$; 1 for $x = 1$; ∞ for $x > 1$.

EXERCISES 3.3

1. (i) $1/n^2$; 0. (ii) -5; -5. (iii) $(-1)^n 2$. (iv) $4-1/2^{n-1}$; 4. (v) $\sqrt{n}$; ∞.
(vi) $1/\sqrt{(2n-1)}$; 0.

2. (i) 0. (ii) ∞. (iii) 3/7.

6. No; e.g. take $a_n = -1/n$.

8. (i) E.g. 1, 1/2, 1/3, 1/4, 1/5. . . . , (iii) E.g. 1, 0, 1/2, 0, 1/4, 0, 1/8, 0. . . . ,
(ii) E.g. 1, 1, 1, 1, 1, 1, 1. . . . (iv) E.g. 1, 1, 1, 1/2, 3/4, 7/8, 15/16. . .

10. (i) E.g. $a_n = 2n$, $b_n = -n$. (iii) E.g. $a_n = n+1$, $b_n = -n$.
(ii) E.g. $a_n = n$, $b_n = -2n$. (iv) E.g. $a_n = n+(-1)^n$, $b_n = -n$.

11. (i) E.g. $a_n = 1/n$, $b_n = 1/n^2$. (iii) E.g. $a_n = b_n = 1/n$.
(ii) E.g. $a_n = 1/n$, $b_n = -1/n^2$. (iv) E.g. $a_n = 1/n$, $b_n = (-1)^n/n$.

12. (i) Yes; e.g. $a_n = b_n = (-1)^n$. (ii) Yes; e.g. $a_n = (-1)^n$, $b_n = 1/n$.

13. (i) False; e.g. $\{4-1/2^n\}$.
(ii) False; e.g. 3, 1, 4, 2, 5, 3, 6, 4. . . (iii) True.

Exercises 3.4

1. (i) E.g. $\{-1/n\}$. (ii) E.g. $\{4+1/n\}$.

3. E.g. 1, -2, 4, -8, 16, -32. . . .

4. (i) E.g. $\{(-1)^n/n\}$. (ii) E.g. $\{(-1)^n\}$.

5. No; e.g. 0, -1, 0, $-\frac{1}{2}$, 0, $-\frac{1}{4}$, 0. . . .

Exercises 3.5

1. (i) 2. (ii) 2. (iii) ∞. (iv) $-\infty$.

2. 1; ∞.

3. $(1+\sqrt{5})/2$.

Exercises 3.6

1. (i) to (vi) are convergent with limits 0,0, 0, 1, 0, 0 respectively. (vii) oscillates and (viii) tends to infinity.

2. (i) and (ii) tend to zero. (iii) oscillates.

Exercises 3.7

1. (i) 3, -5. (ii) 1, -1. (iii) ∞, ∞. (iv) ∞, $-\infty$. (v) 1, -1. (vi) ∞, 1.

4. E.g. $\{a+1/n : n$ is a natural number, a is 1, 2, 3, or 4$\}$.

5. E.g. $\{a+1/n : m$ and n are natural numbers$\}$.

6. E.g. $\{x : |x| \leqslant 1\}$.

7. E.g. the set of rationals.

11. E.g. $a_n = (-1)^n$, $b_n = (-1)^{n+1}$.

12. E.g. 0, 1, 1/2, 1/3, 2/3, 1/4, 2/4, 3/4, 1/5, 2/5, 3/5, 4/5. . . .

Exercises for Chapter 3

5. $\max(a_1, a_2, \ldots, a_n)$. *Hint:* take out $\max(a_1, a_2, \ldots, a_n)$ as a factor.

6. (a) Sequence has limit 1; $n = 8$.

(b) Sequence has no limit.

(c) Sequence has limit 2; $n = 8$. (See section 4.1.)

7. (i) See Theorems 3.3.8 and 3.3.9. (ii) Theorem 3.6.1.

8. See section 2.4. The rationals are not complete; e.g. the set $\{r : r \text{ is a rational}, r < \sqrt{2}\}$ is bounded above but has no least rational upper bound. E.g. any sequence which converges to $\sqrt{5}$.

Exercises 4.1

1. (i) 590. (ii) 3090. (iii) $5/4(1-(1/5)^{20})$. (iv) $4^{20}-1$. (v) $4/3(1-2^{20})$. (vi) 0.

4. $63\frac{1}{2}$.

Exercises 4.2

1. (i), (iii), and (vi) are divergent. (ii) (iv) and (v) are convergent with sums 200, $2\frac{2}{3}$ and zero respectively.

Exercises 4.3

1. (i) Convergent. (ii) Divergent.

2. (i) $\frac{3}{4}$ (ii) $\frac{1}{4}$. (iii) $\frac{1}{4}$.

Exercises 4.4

1. (i), (ii), (iv), (vi), and (vii) are convergent.

Exercises 4.5

2. There are 7–1 = 6 digits in the recurring pattern of 1/7, only 2 in the pattern of 1/11.

3. 0.793. . . .

5. The number of digits in the recurring pattern of the decimal presentation of p/q is a factor of $q-1$, when q is prime. This may be proved using Fermat's theorem, that $10^{q-1} \equiv 1 \pmod{q}$. It is necessary to show that, if r is the least natural number with $10^r \equiv 1 \pmod{q}$, then r is a factor of $q-1$.

EXERCISES 4.6

2. (i) $|x| < 1$ and $x = -1$.
(ii) $|x| \leqslant 1$.
(iii) $|x| < 1$.
(iv) $|x| > 1$.
(v) $x = 0$.
(vi) $|x| < 1$.

3. E.g. $1/2+1/2+1/4+1/4+1/8+1/8+\ldots$.

4. Circular argument. The convergence of $\sum x^n$ is used to prove D'Alembert's test.

EXERCISES 4.7

1. Signs do not alternate strictly.

2. It is absolutely convergent.

3. (i) Absolutely. (ii) Divergent. (iii) Conditionally. (iv) Conditionally.

EXERCISES 4.8

1. 2/3.

4. No effect either on its convergence or on its sum.

EXERCISES 4.9

1. $|x| < 1$; $1/(1-x)\cdot x/(1-x^2)$.

EXERCISES FOR CHAPTER 4.

1. 1024; 2^{118}.

2. *Hint:* use D'Alembert's test and then Theorem 4.2.4.

3. D'Alembert's test; alternating series test; harmonic series.

4. (a) $|x| < 1$ and $x = -1$.
(b) all x.

5. (i) (a) E.g. $\{r : r \text{ is a rational and } |r| < \sqrt{2}\}$.
(ii) (b) meaningless if $T = 0$.
(iii) (b) E.g. harmonic series.
(iv) (b) E.g. $u_n = 1/n^2$.
(v) (b) E.g. $a_n = 1/n$, $b_n = (-1)^n/n$.

6. The second and fourth series are convergent. The convergence of the second may be deduced from the alternating series test.

7. E.g. $a_n = (-1)^n/\sqrt{n}$.

8. (a) E.g. $a_n = (-1)^n/\sqrt{n}$. (b) E.g. $a_n = 1/n$.
E.g. $2-2+2-2+2-2+\dots$

9. (i) (a) Theorem 2.4.2. corollary. (b) Theorem 3.7.9.
E.g. $\{x: 0 < x < 1\}$; $\{1/n : n$ is a natural number$\}$.
(ii) $N = 9$, but rather dependent upon the precise formulation chosen for the definition.

10. (a) $|x| < 1$. (b) $|x| < 1$ or $x = 1$.

11. (iii) $|x| < 1/|a|$. All values of x.

12. (i) E.g. $\sum(-1)^n/n$. (ii) $|x| > 1$.

13. (i) $c_0 = 1$, $c_n = 0$ when $n \geqslant 1$. (Use the binomial theorem.)
(ii) $2x/(1-x)^3$ (see question 1 of exercises 4.9).

14. (a) (i) Diverges. (ii) Converges.
(c) (i) Converges when $|x| < 2$. (ii) Diverges for all x.

16. Series convergent when $|x| < 1$.

17. (a) False; e.g. harmonic series. (b) True.
Convergent if $x = 0$. Convergent if $|x| > 1$.

18. (i) Divergent. (ii) (a) $|x| > \sqrt{2}$. (b) $0 < |x| < \sqrt{2}$.

Exercises 5.1

2. (ii) $\{x : x \neq 0\}$.

Exercises 5.2

2. (i) 9. (ii) ∞. (iii) no limit. (iv) $-\infty$. (v) 6. (vi) no limit. (vii) no limit.

5. (i) 1. (ii) 0. (iii) no limit. (iv) 0. (v) no limit. (vi) 1. (vii) no limit.

Exercises 5.3

2. (i) 1. (ii) -7.

Exercises 5.4

3. f is continuous (i) everywhere, (ii) for $x \neq -5$, (iii) for $|x| > 1$, (iv) for $|x| \neq 2$, (v) only at zero, (vi) nowhere, (vii) for $|x| \neq 1$.

5. (i) E.g. $f(x) = x^2$.

(ii) E.g. $\begin{cases} f(x) = 0 & \text{if } x \text{ is a rational,} \\ f(x) = 1 & \text{if } x \text{ is an irrational.} \end{cases}$

(iii) E.g. $\begin{cases} f(x) = x & \text{if } x \text{ is a rational,} \\ f(x) = 0 & \text{if } x \text{ is an irrational.} \end{cases}$

(iv) E.g. $\begin{cases} f(x) = x(x-1)(x-2)(x-3) & \text{if } x \text{ is a rational,} \\ f(x) = 0 & \text{if } x \text{ is an irrational.} \end{cases}$

(v) E.g. $f(x) = 1/(x-1)(x-2)$.

(vi) E.g. $f(x) = [x]$.

EXERCISES 5.5

1-6 are all false. The reader is left to provide counterexamples.

EXERCISES 5.6

2. Yes.

3. E.g. $f(x) = \cos 1/x(x-1)$.

4. Removable discontinuity at $x = 2$; non-removable at $x = -2$.

5. E.g. $\begin{cases} f(x) = 5 & \text{if } x \leqslant 3, \\ f(x) = 2 & \text{if } x > 3. \end{cases}$

6. E.g. $\begin{cases} f(x) = 5 & \text{if } x \neq 2, \\ f(2) = 2. \end{cases}$

7. E.g. $\begin{cases} f(x) = 0 & \text{if } x \text{ is a rational or if } |x| < 1, \\ f(x) = 1 & \text{if } x \text{ is an irrational and } |x| > 1. \end{cases}$

8. (i) and (iii) are.

10. The following have jump discontinuities: (ii) at the integers, (iii) at zero, (iv) at zero and 1.

EXERCISES 5.7

2. E.g. f is defined in the closed interval [0, 1] as follows:

$$\begin{cases} f(x) = 1 & \text{if } x \leqslant \frac{1}{2}, \\ f(x) = 0 & \text{if } x > \frac{1}{2}. \end{cases}$$

3. No.

4. E.g. $f(x) = 2$.

5. E.g. $\begin{cases} f(x) = x & \text{if } 0 < x < 1, \\ f(0) = f(1) = \frac{1}{2}. \end{cases}$

6. (i) 4, 0. (ii) 3, 0. (iii) $\pi/2$, 4.

7. E.g. in [0, 2] $\begin{cases} f(x) = x & \text{if } 0 \leqslant x \leqslant 1 \\ f(x) = \frac{1}{2} & \text{if } 1 < x \leqslant 2. \end{cases}$

Exercises for Chapter 5

3. (i) (b) 1. (c) 1 (see Theorem 7.4.2).
(ii) (a) $m = -\frac{1}{4}$, $M = 2$. (b) 2; $1\frac{1}{2}$.

Exercises 6.1

3. 12.

5. (i) -1. (ii) -1, 0.

6. (i) $3x^2 \cos x - x^3 \sin x$.
(ii) $\cos x/x - \sin x/x^2$.
(iii) $(-1/x^2) \cos 1/x$.
(iv) $-3 \cos^2 x \sin x$.
(v) $6 \sin 3x \cos 3x$.
(vi) $3x^2 \tan^2 x + 2x^3 \tan x \sec^2 x$.

Exercises 6.2

3. $u = 2$.

4. Use Darboux's theorem.

Exercises 6.3

1. $u = 1$.

3. $u = -\sqrt{10}$.

7. Because $f(0) = 0$.

Exercises 6.4

1. Twice.

4. Once, since the first derivative is defined only at this point.

5. E.g. $f(x) = x^3$ in the interval $(-1, 1)$.

6. (i) $1/\sqrt{3}$; both. (ii) no minima. (iii) 0; (a). (iv) 0; (a). (v) 0; neither.

7. First form, since $f'(x)$ is defined only at $x = 0$.

10. (i) 3/5, (ii) $-3/2$ (iii) 1/6. (iv) $-1/8$. (v) $-1/2$. (vi) $-1/4$.

EXERCISES 6.5

2. $1-x+x^2-x^3+x^4-x^5+\ldots(-x)^n+\ldots.$

3. $1+4x+12x^2+32x^3+\ldots+n(2x)^n+\ldots$

4. $\sqrt{3}/2-(1/2)x-(\sqrt{3}/4)x^2+(1/12)x^3+\ldots.$ 0.85

EXERCISES FOR CHAPTER 6.

1. *Hint:* the proof of (i) by induction depends upon showing that if $a_{n+1} > a$ and $2a_n^3-16a_n+15 > 0$, then $2a_{n+1}^3-16a_{n+1}+15 > 0$.

2. (i) Neither. (ii) Continuous. (iii) Neither.

3. $3x|x|$. f'' exists at $x = 0$, but not f'''.

4. Construct the function $\{f(x)-f(a)\}\{g(b)-g(x)\}$.

5. (a) not differentiable at 1. (b) not continuous at zero. $(a+b)/2$.

6. $k = 4$.

7. 0.777.

8. (a) 0.527.
(b) This is the first form of L'Hopital's rule to be proved by the method of the second form.

10. Make $f(x) = \cos 3x+\cos 2x+\cos x$.

11. Theorem 6.2.3 and 6.2.4. Define $g(0)$ to be the greatest lower bound of g in $(0, 1]$.

12. $a_{2n} = 0$; $a_{2n+1} = (-1)^{n+1}(2^{2n+1})/(2n+1)!$; $b_n = (-1)^n(n+1)4^{n+2}/\pi^{n+2}$. All x. $0 < x < \pi/2$. sinh 2. $1/(\pi/2-x)^2$.

13. *Hint:* see exercises for Chapter 3, question 2.

14. $\alpha-1$, 1. $1+\alpha^2/8$, 1. $1+\alpha^2/8$, $\alpha-1$.

EXERCISES 7.1

1. (i) $R = 1$.
(ii) $R = 0$.
(iii) $R = 3$; convergent when $x = -3$.
(iv) $R = \frac{1}{4}$; convergent when $x = \frac{1}{4}$ or $x = -\frac{1}{4}$.
(v) $R = 1$; convergent when $x = 1$.

2. 1, $1/(1-x)$, 1.

EXERCISES 7.2

2. Minimum when $x = 0$; maximum when $x = -2$.

3. All limits are zero.

4. (i) Yes. (ii) No.

9. 1.

10. $\sum_{n=0}^{\infty} \frac{x^{2n}}{(2n)!}$. $\sum_{n=0}^{\infty} \frac{x^{2n+1}}{(2n+1)!}$.

EXERCISES 7.3

3. (i) 1. (ii) 1. (iii) $-\frac{1}{2}$. Removable.

4. $1+x-x^3/3$.

EXERCISES 7.4

1. (i) 2. (ii) 1/3 log 2.

3, 4. 0.6931; 1.099; 1.792; 5.118

5. 0.3010.

7. Zero.

8. (i) $(\cos x \log x+\sin x/x)x^{\sin x}$. $\{x : x > 0\}$.

(ii) $2^x \log 2$. All x.

(iii) $[\log (\sin x)+x \cot x] (\sin x)^x$. $\{x: 2n\pi < x < (2n+1)\pi\}$.

(iv) $[(\log x+1) \log x \cdot x^x+x^{x-1}]x^{x^x}\{x : x > 0\}$.

EXERCISES 7.6

1. $(a-n)x/n$.

2. See exercises 7.4, question 5.

EXERCISES FOR CHAPTER 7.

1. $f(0) = 1; f'(0) = -1. -1 < f(x) < 1$ if $x > 0$.

2. See question 9 of exercises 7.4.

3. (a) 1. (b) $\frac{1}{2}$.

4. See Theorem 7.4.6. $\sum_{n=1}^{\infty} (-1)^{n+1} x^{2n}/n \; |x| \leqslant 1$.

6. (i) Convergent when $|x| < 1$; divergent when $|x| > 1$.
(ii) (a) Divergent for all x. (b) Convergent if $|x| < 1$.

7. $x(1+x)/(1-x)^3$.

8. Diverges.

10. -1.

EXERCISES 8.2

4. $h^4/4$. $173\frac{1}{4}$.

5. (i) $12\frac{1}{2}$. (ii) 10. (iii) $-2\frac{1}{2}$.

10. 256.

EXERCISES 8.3

2. E.g. $f(x) = \sinh x$, $f(x) = \cosh x$, $f(x) = \log x$.

EXERCISES 8.4

2. Suppose f is discontinuous at c. Then f is integrable in $[a, c-\delta]$ and $[c+\delta, b]$ The result follows, by letting δ tend to zero and using Theorem 8.4.1.

3. E.g. $\begin{cases} f(x) = 1/x & \text{if } x \neq 0, \\ f(0) = 0. \end{cases}$

4. (i) $\begin{cases} f(x) = x^2/2 & \text{if } x \geqslant 0, \\ f(x) = -x^2/2 & \text{if } x < 0. \end{cases}$

(ii) $\begin{cases} f(x) = x^2/2 & \text{if } x \leqslant 1, \\ f(x) = x^3/3+1/6 & \text{if } x > 1. \end{cases}$

(iii) $\begin{cases} f(x) = x^2/2-2x & \text{if } x \geqslant 2, \\ f(x) = 2x-x^2/2-4 & \text{if } x < 2. \end{cases}$

(iv) $\begin{cases} f(x) = x^2 \log |x| - x^2/2 & \text{if } x \neq 0, \\ f(0) = 0. \end{cases}$

5. (i) 0. (ii) Not. (iii) 7/12.

6. E.g. in [0, 1] let f and g be defined by

$$\begin{cases} f(x) = 1 \text{ if } x \text{ is a rational,} \\ f(x) = 0 \text{ if } x \text{ is an irrational,} \end{cases}$$

$$\begin{cases} g(x) = 0 \text{ if } x \text{ is a rational,} \\ g(x) = 1 \text{ if } x \text{ is an irrational.} \end{cases}$$

7. $-0.29 < \int_{\pi/6}^{\pi/2} \log \sin x \, dx < -0.16.$

EXERCISES 8.5

1. 3.142
3. $x+x^3/6+3x^5/40+\ldots$
4. $2/\pi$.

EXERCISES 8.6

1. 3.76; -1.02
2. Nonsense. $1/x$ is not continuous at zero.
6. $n \log n-n+1$. Zero.
7. $\log 2+R_{2n}-R_n$; $\log 2$.
8. $\frac{1}{2} \log 8+R_{4n}-\frac{1}{2}R_{2n}-\frac{1}{2}R_n$.
9. $\frac{1}{2} \log 4n+R_{2n^2}-\frac{1}{2}R_{n^2}-\frac{1}{2}R_n$.
10. 10^{14}.

EXERCISES 8.7

1. (i) Divergent. (ii) 1. (iii) Divergent. (iv) Divergent. (v) 1. (vi) 0.
3. Use Theorem 8.7.1.

EXERCISES FOR CHAPTER 8.

1. Fundamental theorem of calculus. Zero.
2. $-2/\pi^2$.
3. Not bounded. Upper and lower integrals unequal. Continuous.
5. (ii) $s_n = t_n = (-1)^n$. (iii) See Theorem 8.7.1.
10. π.
13. (i) e^a. (ii) $1/\sqrt{e}$.

INDEX